SpringerBriefs in Food, Health, and Nutrition

Springer Briefs in Food, Health, and Nutrition present concise summaries of cutting edge research and practical applications across a wide range of topics related to the field of food science, including its impact and relationship to health and nutrition. Subjects include: Food Chemistry, including analytical methods; ingredient functionality; physic-chemical aspects; thermodynamics Food Microbiology, including food safety; fermentation; foodborne pathogens; detection methods Food Process Engineering, including unit operations; mass transfer; heating, chilling and freezing; thermal and non-thermal processing, new technologies Food Physics, including material science; rheology, chewing/mastication Food Policy And applications to: Sensory Science Packaging Food Qualtiy Product Development We are especially interested in how these areas impact or are related to health and nutrition. Featuring compact volumes of 50 to 125 pages, the series covers a range of content from professional to academic. Typical topics might include:

- A timely report of state-of-the art analytical techniques
- A bridge between new research results, as published in journal articles, and a contextual literature review
- A snapshot of a hot or emerging topic
- An in-depth case study
- A presentation of core concepts that students must understand in order to make independent contributions

Bee Ling Tan • Mohd Esa Norhaizan

Rice Noodles: Quality Improvement, Processing Techniques and Nutritional Aspects

Springer

Bee Ling Tan
Department of Diagnostic
and Allied Health Science
Faculty of Health and Life Sciences
Management and Science University
Shah Alam, Selangor, Malaysia

Mohd Esa Norhaizan
Department of Nutrition
Faculty of Medicine
and Health Sciences
Universiti Putra Malaysia
Serdang, Selangor, Malaysia

ISSN 2197-571X ISSN 2197-5728 (electronic)
SpringerBriefs in Food, Health, and Nutrition
ISBN 978-3-031-98648-2 ISBN 978-3-031-98649-9 (eBook)
https://doi.org/10.1007/978-3-031-98649-9

This Springer imprint is published by the registered company Springer Nature Switzerland AG
The registered company address is: Gewerbestrasse 11, 6330 Cham, Switzerland

Preface

Rice (*Oryza sativa* L.) is a fundamental principal food for more than half of the world's people, providing more than 20% of the calories consumed globally, particularly in Latin America and West Indies, Middle East, and South and East Asia. Rice is usually consumed as fully mature grains in the form of whole kernels after boiling in excess water or cooking using a certain amount of water. Besides traditional consumption, rice has been increasingly used as an ingredient of various food formulations including rice-based products. Rice-based food products are commonly consumed globally, providing up to 43% and 46% of dietary protein and dietary energy in the sub-Saharan region, respectively. Rice noodles also known as rice noodle strips or rice vermicelli, based on the rice flour or rice starch obtained by wet-milling, are one of the popular types of rice-based products consumed in Asian countries. There are various sizes and shapes of rice noodles, namely, short and long, round and flat, and thin and thick. Traditionally, rice noodles are made from long-grain rice with intermediate to high amylose content. Since amylose networks play a crucial role in the structure of rice noodles, high amylose rice is often utilized for making rice noodles. Nonetheless, rice noodles made from different rice cultivars give different qualities. Amylose plays a crucial role in regulating the glycemic index. Starchy foods that are rich in amylose are linked to slower emptying of the human gastrointestinal tract and lower blood glucose levels compared to those with low levels of amylose. Indeed, rice noodles are the potential rice-based products that can be formulated to be a lower glycemic index value compared to other rice products. Yet, mixed cultivars of rice flour are usually used to produce rice noodles in industry and thus leading to high variation in noodle qualities. Furthermore, the shelf life and quality of rice noodles are also affected by the manufacturing process and storage conditions, particularly fresh rice noodles. Based on our knowledge, the literature reported on rice noodles as well as their nutritional aspect, quality improvement, and processing technique has not well been compiled in the form of a brief/book. Indeed, this disintegrated literature needs to be compiled together to deliver information at one point. Hence, this book attempts to discuss several issues associated with rice noodles, namely, classification of noodles, processing of rice noodles, raw materials and their roles, factors affecting the qualities of rice noodles, qualities and nutritional values improvement of rice noodles, the glycemic index of the rice noodles, and future prospects. By summarizing all the literature cohesively and lucidly in one book/brief, it would provide a cohesive representation of the information and practical reference on the processing and production of rice noodles, highlighting their factors essential for the quality of rice noodles as well as their nutritional values and glycemic index of rice noodles, which would be interesting for the readers and allied stakeholders.

Shah Alam, Selangor, Malaysia Bee Ling Tan

Contents

Chapter 1
Background and Introduction

Contents

Rice is a staple food that feeds over half of the global population. Rice is a predominant source of calories that provides vitamins (folic acid, niacin, and thiamine) and minerals (phosphorus, iron, and selenium) (Fukagawa and Ziska 2019; Kasote et al. 2022). Besides vitamins and minerals, it also contains phytochemicals including flavonoids, anthocyanins, and polyphenols (Melini et al. 2019; Bagchi et al. 2021). Rice is cultivated in over 100 countries with 90% of the world's production from Asian countries (Priya et al. 2019; Food and Agriculture Organization Corporate Statistical Database (FAOSTAT) 2020). The production of paddy rice in several countries from 2000 to 2019 is shown in Table 1.1. Despite a lack of historical records that might affect the precise determination, research evidence strongly suggests that rice originated in southeast continental Asia (Houston and Kohler 1970). It has been revealed that there are more than 110,000 varieties of rice, in which *Oryza sativa* is the most commonly grown rice. Indeed, the production practices/growing conditions, aroma, stickiness, thickness, color, and grain length of rice may affect the nutrients and quality profiles of rice (Zhou et al. 2018; Rathna et al. 2019; Ren et al. 2020; Carcea 2021; Li et al. 2021).

Rice is a semi-aquatic annual grass plant including 22 species of the genus *Oryza*, in which 20 of them belong to the wild types (Singh et al. 2021). In general, two species of rice are crucial for human consumption, namely *O. glaberrima* and *O. sativa* (Bharamappanavara et al. 2020). Rice is primarily consumed as white rice, a refined grain that is produced from the rice milling process which removes the germ and bran and leaves the starchy endosperm. Rice has many unique characteristics such as hypoallergenic properties, a mild flavor, and ease of digestion (Abid et al. 2024).

B. L. Tan, M. E. Norhaizan, *Rice Noodles: Quality Improvement, Processing Techniques and Nutritional Aspects*, SpringerBriefs in Food, Health, and Nutrition, https://doi.org/10.1007/978-3-031-98649-9_1

Table 1.1 The production of paddy rice in several countries from 2000 to 2019

	Production quantity (Million tonnes)				
Country	2000	2005	2010	2015	2019
Africa	17.4	20.2	25.5	29.8	38.7
America	31.6	36.2	36.5	37.2	35.3
Asia	545.2	574.0	627.4	659.9	677.2
India	127.4	137.6	143.9	156.5	177.6
China	189.8	182.0	197.2	213.7	211.4
Bangladesh	37.6	39.7	50.0	51.8	54.5
Indonesia	51.8	54.1	59.2	61.0	54.6
Malaysia	2.1	2.3	2.4	2.7	2.9
Europe	3.1	3.3	4.3	4.2	4.0
World	598.6	634.2	694.0	731.9	755.4

Source: Food and Agriculture Organization of the United Nations (FAO) (2021)

Rice noodles (Table 1.2) are traditional foods commonly consumed in Southeast Asia (Pichaiyongvongdee et al. 2022). The primary rice variety used for rice noodles is *Indica*, which is widely cultivated in South China due to its high adaptability and yield in local climates (Zhang et al. 2024). *Indica* rice, with its higher amylose content, is more suitable for rice noodle production compared to *Japonica* rice (Lu et al. 2008). Rice noodles, which originated during the Qin dynasty (259–210 B.C.), have been a staple food in China for over 2000 years. These noodles are typically served as a side dish or main food, often mixed with vegetables and meats for stir-frying or boiled in broth to create soup noodles (Fu 2008).

Flour plays an essential role in noodle quality. In general, rice noodles are made from long-grain rice with intermediate to high amylose content (more than 22% amylose) that contributes to the gel network. The degree of pre-gelatinization of flour plays a crucial role in imparting desirable noodle texture. Despite a certain degree of gelatinization needed to produce the binding effect during extrusion, excessive gelatinization may lead to extremely high extrusion pressure (Qiu et al. 2024). In addition, the protein content is crucial for the textural characteristics. Indeed, adequate gluten extensibility and strength are needed in noodle flour (Zang et al. 2022). Rice protein lacks the functionality of wheat gluten in making a cohesive dough structure. Subsequently, at least part of the rice flour is pre-gelatinized to create a binder for the remaining flour to create a uniform matrix for embedding the starch granules (Yao et al. 2023).

There are several key procedures in processing raw rice into rice noodles. The steps consist of selecting raw rice, cleaning, soaking ground rice for a few hours, grinding, steam cooking rice slurry to gelatinize the rice starch, kneading the slurry to obtain a cohesive dough, extruding the dough, lastly subjecting the extruded noodles to surface gelatinization in boiling water to enhance the noodles texture and stability, and drying (Juliano and Sakurai 1985; Li et al. 2015). Despite the production procedure used for rice noodles may differ from region to region to

Table 1.2 Classification of rice noodles based on molding method

Type of noodle	Molding	Shape	Description of rice noodles	Moisture content[a] (%)	Final products
Zhafen	Extruding	Round	Length: 200–400 mm	40–50	*Changde-mifen*
			Diameter: 2–3 mm		
			(Fermented *mifen*)		
			Length: 200–400 mm	40–50	*Guilin-mifen*
			Diameter: 2–3 mm		
			Length: 50 mm	Low: 15–25	*Fenli*
			Diameter: 20 mm	High: 40–50	
			Length: 200–400 mm	Low: 15–25	*Guoqiao-mixian*
			Diameter: 1–2 mm	High: 40–50	
Qiefen	Cutting	Rectangle	Length: 200 mm	Low: 15–25	*Shahe-fen, qiefen, bianfen*
			Width: 4–6 mm	High: 40–65	
			Thickness: 1 mm		
Others	Rolling and spreading	Irregular	Width: 20–30 mm	55–65	*Changfen*
			Thickness: 1 mm		
			(Stuffing is rolled in it)		

Source: Li et al. (2015)
[a]Low or high: distinguishes whether the rice noodles are dehydrated (low) or high moisture

meet local needs, the principles are generally the same. Sheeted and flat fresh rice noodles are famous in many regions of Southeast Asia, Japan, and Southern China (Ul Haque et al. 2022). The dough is placed in a rotating heated drum to form a sheet. The sheet is then placed in a steamer thus leading to a partial gelatinization of starch. The sheets (about 1 mm thick) are passed through the cutting rolls to form noodle strands. The noodles are then dried or served as fresh rice noodles (Yang et al. 2024).

According to the best of the authors' knowledge, the literature on rice noodles and their processing technique, quality improvement, and nutritional aspects has not well been compiled in the form of a book/brief. Indeed, this disintegrated information needs to be compiled together to deliver knowledge at one point. Therefore, this book explores several issues linked to rice noodles, namely types of noodles, processing of rice noodles, raw materials and their roles, factors affecting the qualities of rice noodles, qualities and nutritional values improvement of rice noodles, the glycemic index of the rice noodles, and future prospects.

References

Abid J, Ahmed S, Xia T et al (2024) Rice as a nutritional grain: examining its role in healthy products and disease prevention. Food Rev Int 40:3055–3078

Bagchi TB, Chattopadhyay K, Sivashankari M et al (2021) Effect of different processing technologies on phenolic acids, flavonoids and other antioxidants content in pigmented rice. J Cereal Sci 100:103263

Bharamappanavara M, Siddaiah AM, Ponnuvel S et al (2020) Mapping QTL hotspots associated with weed competitive traits in backcross population derived from *Oryza sativa* L. and *O. glaberrima* Steud. Sci Rep 10:22103

Carcea M (2021) Value of whole grain rice in a healthy human nutrition. Agriculture 11:720

Food and Agriculture Organization Corporate Statistical Database (FAOSTAT) (2020) Food and Agriculture Organization of the United Nations Database. Food and Agriculture Organization (FAO), Rome. http://www.fao.org. Accessed 12 May 2020

Food and Agriculture Organization of the United Nations (FAO) (2021) FAOSTAT crops. http://www.fao.org/faostat/en/#data/QC. Accessed 6 July 2021

Fu BX (2008) Asian noodles: history, classification, raw materials, and processing. Food Res Int 41:888–902

Fukagawa NK, Ziska LH (2019) Rice: importance for global nutrition. J Nutr Sci Vitaminol 65:S2–S3

Houston DF, Kohler GO (1970) Chapter 1: Background and present situation. In: Nutritional properties of rice. National Academy of Sciences, Washington, DC, pp 1–7

Juliano BO, Sakurai J (1985) Miscellaneous rice products. In: Juliano BO (ed) Rice: chemistry and technology, 2nd edn. Am Assoc Cereal Chem, St. Paul, pp 569–618

Kasote D, Sreenivasulu N, Acuin C et al (2022) Enhancing health benefits of milled rice: current status and future perspectives. Crit Rev Food Sci Nutr 62:8099–8119

Li Y, Liang J-F, Yang M-Y et al (2015) Traditional chinese rice noodles: history, classification, and processing methods. Cereal Foods World 60:123–127

Li H, Xu M, Chen Z et al (2021) Effects of the degree of milling on starch leaching characteristics and its relation to rice stickiness. J Cereal Sci 98:103163

Lu Z-H, Cao W, Peng H-H et al (2008) Effect of fermentation metabolites on rheological and sensory properties of fermented rice noodles. J Sci Food Agric 88:2134–2141

Melini V, Panfili G, Fratianni A et al (2019) Bioactive compounds in rice on Italian market: pigmented varieties as a source of carotenoids, total phenolic compounds and anthocyanins, before and after cooking. Food Chem 277:119–127

Pichaiyongvongdee S, Foophow T, Rasamipaiboon N et al (2022) Effects of different particle size distribution and insoluble dietary fiber content from Pomelo by-products on the quality characteristics of rice noodle products. J Food Health Bioenviron Sci 15:13–25

Priya TSR, Nelson ARLE, Ravichandran K et al (2019) Nutritional and functional properties of coloured rice varieties of South India: a review. J Ethn Foods 6:1–11

Qiu C, Hu H, Chen B et al (2024) Research progress on the physicochemical properties of starch-based foods by extrusion processing. Foods 13:3677

Rathna PTS, Nelson ARLE, Ravichandran K et al (2019) Nutritional and functional properties of coloured rice varieties of South India: a review. J Ethn Foods 6:11

Ren C, Hong B, Zheng X et al (2020) Improvement of germinated brown rice quality with autoclaving treatment. Food Sci Nutr 8:1709–1717

Singh VJ, Vinod KK, Krishnan SG et al (2021) Chapter 1: Rice adaptation to climate change: opportunities and priorities in molecular breeding. In: Hossain MA, Hassan L, Ifterkharuddaula KM et al (eds) Molecular breeding for rice abiotic stress tolerance and nutritional quality. John Wiley & Sons, Ltd, Hoboken, p 436

Ul Haque E, Afzaal S, Hayat A et al (2022) Rice-based products. In: Sarwar N, Atique-ur-Rehman, Ahmad S, Hasanuzzaman M (eds) Modern techniques of rice crop production. Springer, Singapore, pp 781–807

Yang Z, Liu P, Fang X et al (2024) The selection of storage period for fresh rice noodles, processing materials, and the screening of key indicators. Foods 13:3965

Yao Y, Zhou C, Wang J et al (2023) Improving of noodle quality caused by starch-protein interaction of waxy and strong-gluten wheat flour. LWT 188:115394

Zang P, Gao Y, Chen P et al (2022) Recent advances in the study of wheat protein and other food components affecting the gluten network and the properties of noodles. Foods 11:3824

Zhang J, You Y, Li C et al (2024) The modulatory roles and regulatory strategy of starch in the textural and rehydration attributes of dried noodle products. Crit Rev Food Sci Nutr 64:5551–5567

Zhou C, Huang Y, Jia B et al (2018) Effects of cultivar, nitrogen rate, and planting density on rice-grain quality. Agronomy 8:246

Chapter 2
Types of Noodles

Contents

Noodles have become a crucial dietary component in the human diet (Ahmed et al. 2024). While there are contentious issues surrounding where noodles are originated (Donadio 2009), the discovery of a pot of well-preserved 4000 years old noodles unearthed in 2005 by Chinese archaeologists in the Lajia archaeological site in northwestern China (Lu et al. 2005). In this regard, these easily recognizable noodles are more than 2000 years older than the earliest mention of noodles, which is appeared in a Chinese book written during the East Han Dynasty between 25 and 220 AD. The noodles were yellow, more than 50 cm in length, and thin (~0.3 cm in diameter). It resembles the *La-Mian* noodle, a traditional Chinese noodle that is made by repeatedly stretching and pulling the dough by hand. These turned out that these 4000 years old noodles were made from millet, and not from wheat flour as they are commonly available today (Hou 2010).

In the Eastern Han Dynasty (25–220 AD), Chinese noodles were introduced to Japan by the Japanese envoy to China. Subsequently, noodles were spread from China to other Asian countries and beyond, where they were adapted into the local diet and modified, ultimately evolving into various forms and preparations that became an essential part of local cuisines. During the thirteenth century, Marco Polo

B. L. Tan, M. E. Norhaizan, *Rice Noodles: Quality Improvement, Processing Techniques and Nutritional Aspects*, SpringerBriefs in Food, Health, and Nutrition,
https://doi.org/10.1007/978-3-031-98649-9_2

brought Chinese noodle-making technology back to Europe, where noodles were evolved into pasta products (Hou 2001). Despite certain Italian pasta (such as spaghetti) and Asian noodles appearing to be almost similar in shape, there are some differences between both pasta and Asian noodles in terms of their consumption patterns, the process involved, raw materials used, and characteristics (Hou 2001). Noodles are characterized by thin strips slit from a sheeted dough that are primarily made from wheat flour (*Triticum aestivum*), salt (alkaline salt or common salt), and water. It can be served in different shapes and forms depending on the preferences of the community. In general, noodles are categorized based on the form of noodles, size of noodle strands, salt composition, processing methods, and raw materials (Meenu et al. 2023). Noodles are classified into thin handmade noodles (tenobe somen) and flat or handmade standard noodles (teuchihira-men or teuchiudon) based on the manufacturing process. Noodles are further classified into machine-made noodles, instant cup noodles, frozen boiled noodles, steamed noodles (mushi-men), dried noodles (kan-men), boiled noodles (yude-men), and instant noodles (Fu 2008). In addition, they are some other noodles including pasta, rice noodles, starch noodles, Naengmyon noodles (Korean type noodles), Chinese type wheat noodles, buck wheat noodles (Soba), and Japanese type wheat noodles (Nagao 1996; Crosbie et al. 1999). Noodle made from wheat flour remains the most popular noodle product in Asia and around the world, followed by rice and starch-based noodles, which are consumed predominantly in Asia. Noodles are usually consumed in a soup. Pasta is a dried product traditionally made from durum wheat (Triticum durum) semolina and water. The semolina, a coarse durum wheat flour, is mixed with water to form a dough. This dough is then extruded through metal dies under pressure, shaping it into various pasta forms (Mahmud et al. 2023).

2.1 Classification of Noodles

Noodles are a staple food in many Asian countries due to their convenience, affordability, and ease of preparation, contributing to their growing global popularity (Mahmud et al. 2023). Approximately 20%–50% of the total wheat flour consumption in Asia is attributed to noodle production. Although traditionally a significant part of Asian diets, noodles have gained widespread acceptance across the globe. Noodles are typically made from various raw materials, including wheat flour, buckwheat flour, rice flour, and starches derived from corn, sago, sweet potato, tapioca, mung bean, and rice. Wheat flour-based noodles remain the most widely consumed type worldwide, particularly in Asia, followed by rice and starch-based noodles, which are primarily popular in Asia. The diversity of noodles across countries and regions can be attributed to variations in local culture, regional preferences, and climatic conditions, which influence the ingredients and preparation methods (Obadi et al. 2022a). Table 2.1 summarizes the classification of Asian wheat noodles based on processing.

Table 2.1 Classification of Asian wheat noodles based on processing

Type of noodle	Processing
Fresh raw	Noodles emerging from slitting rolls are cut into predetermined lengths and packaged without additional processing. Examples include soba (buckwheat) noodles, Thai bamee (egg noodles), chuka-men, Japanese noodles, and Chinese raw noodles. These noodles are best consumed within 24 h of production due to their tendency to discolor. However, refrigeration can extend their shelf life to 3–5 days, allowing for short-term storage while preserving their freshness and quality.
Air dried	Fresh raw noodles are usually dried in a relative humidity and controlled temperature chamber. Soba noodles, Japanese noodles, chuka-men, and Chinese raw noodles can be in dried form. The shelf life of noodle is extended from several months to 1 year, but handling dried noodles can be challenging due to their fragile nature, which makes them prone to breakage during packaging, transport, and use.
Parboiled	Fresh raw noodles are parboiled (about 70%–80% of complete cooking). Examples include Japanese noodles, chuka-men, and Chinese alkaline noodles. After parboiling, Chinese alkaline noodles (Hokkien) are rinsed in cold water, drained, and coated with 1%–2% vegetable oil to prevent sticking. The boiled Japanese noodles and chuka-men are not mixed with any oil. They are recooked for another 1–2 min before serving.
Steamed	Fresh raw alkaline noodles are steamed in a steamer and softened with water via steeping or rinsing, which is known as "Yakisoba", and it is usually prepared by stir-frying.
Instant	Fresh raw noodle strands are steamed and waved for 2–3 min. After cutting and forming into a block, the noodles are dehydrated by hot blast air (70–80 °C) for 35–45 min or deep-frying in hot oil (135–150 °C) for 1–2 min. The shelf life of instant noodles is 6–12 months. They can be served after soaking or cooking in hot water for 3–4 min.
Frozen	Parboiled or fresh raw noodles are often quickly frozen for 20–30 min until the core temperature drops below −15 °C. Two common types of frozen noodles are produced: frozen boiled noodles and frozen raw noodles, with frozen boiled noodles being more popular. The shelf life of frozen noodles ranges from 6 to 12 months, depending on the storage conditions in the freezer. Before serving, frozen noodles are typically cooked in boiling water for about 1 min.
Long shelf life	Fresh noodles undergo a series of processing steps, including boiling, rinsing, soaking in acidified water, and packaging. These noodles are then subjected to cooling and thermal processing to ensure safety and extend shelf life. Long shelf-life noodles, prepared through this method, can be stored for 4–6 months under appropriate conditions. For consumption, these noodles are rehydrated by soaking in hot water.
Semi-dried	Fresh raw noodles are dried in a chamber with controlled temperature and relative humidity to reduce their moisture content to approximately 25%. This drying process helps extend the shelf life of the noodles to 1–2 months by lowering water activity, provided they are properly packaged. The cooking time for these noodles is shorter compared to fully dried noodles. White salted noodles can also be produced using this method, offering a convenient and longer-lasting option for consumers.

Source: Hou and Kruk (1998) and Hou (2020)

2.2 Raw Materials

Noodles can be made from a mixture of buckwheat flour and hard wheat flour. When wheat flour constitutes less than 40% of the mixture, the resulting noodles are known as soba, which translates to buckwheat noodles. Soba is a popular dish in Korea and Japan (Sadiku et al. 2023). These noodles typically have a grey or light brown color and are characterized by a distinct flavor and taste (Obadi et al. 2024).

Noodles can be classified into two major groups, namely Japanese-type and Chinese-type, based on the type of wheat used. Japanese noodles are typically made from soft wheat flour with medium protein content, resulting in an elastic and soft texture with a creamy white color (Obadi et al. 2022a). In contrast, Chinese-type noodles are made from hard wheat flour, which gives them a firm texture and a bright yellow or creamy white color (Meenu et al. 2023). Other types of noodles may fit into either category, depending on the ingredients and characteristics.

2.3 Salt Used

Noodles can be categorized into two classes based on the presence or absence of alkaline salt such as yellow alkaline noodles (containing alkaline salt) and white salted noodles (containing common salt) (Du et al. 2023). Alkaline salt contributes to the yellow color of the noodles. White salted noodles include Korean white salted noodles, Japanese noodles, and Chinese raw or dry noodles. On the other hand, yellow alkaline noodles include Thai bamee noodles, instant noodles, chuka-men, Cantonese noodles (with or without eggs), and Chinese wet noodles such as Hokkien noodles (Meenu et al. 2023).

2.4 Size

Japanese noodles are classified into four types based on the width of their strands (Table 2.2). Thinner noodles, such as hiya-mugi and so-men, soften more easily in hot water and are typically served cold during the summer months. In contrast, thicker noodles like hira-men and udon are more resistant to softening and are typically consumed hot, making them more popular in cooler seasons. Each type of noodle has a distinct size, which influences its culinary use and seasonal preferences (Obadi et al. 2022b).

Table 2.2 Classification of Japanese noodles according to width

Type of noodles	Size
Hira-men	Flat, 5.0–6.0 mm wide
Udon	Standard, 1.9–3.8 mm wide
Hiya-mugi	Thin, 1.3–1.7 mm wide
So-men	Very thin, 0.7–1.2 mm wide

Source: Nagao (1996)

2.5 Processing

Noodles can be classified based on their method of production, specifically machine-made and handmade varieties. Handmade noodles, which were more common in Asia before the invention of automatic noodle machines in the 1950s, are highly valued for their unique texture. The process of stretching noodles by hand is considered more of an art form than a simple food preparation technique. However, noodle machines are ideal for mass production, ensuring consistent output in larger quantities (Gao et al. 2022). Regardless of the production method, several processing steps are involved, including slitting, rolling/sheeting, compounding, dough sheeting, and mixing raw materials. These procedures are consistent across countries for all noodle types, and the noodle strands are further processed to create different varieties (Zhang et al. 2022).

2.6 Wheat Flour Noodles

Asian noodles have become a staple food in many countries across the region, with noodle consumption accounting for an average of 20%–50% of the total wheat flour intake in these areas (Du et al. 2023). Noodles made from wheat flour can be divided into two main categories based on the presence or absence of alkaline salts: alkaline noodles and regular salted noodles (Obadi and Xu 2021). The fundamental steps in machine-made noodle production such as dough mixing, cutting, sheeting/reduction, compounding, and sheet forming remain largely consistent across different types of noodles (Fu 2008). The preparation of wheat flour noodles typically involves mixing crumbly dough, created by combining one part alkaline salt (*kansui*) and/or sodium chloride solution with three parts of wheat flour. The dough is then compressed between a series of rollers to form a sheet, which is subsequently slit into noodle strands of desired dimensions. These strands are further processed to enhance ease of preparation, modify their eating characteristics, and extend their shelf life. The addition of alkaline salt in some Chinese noodle varieties imparts a distinct alkaline flavor, a firm and elastic texture, and a yellowish color (Hou 2020). Figure 2.1 shows the classification of wheat flour noodles.

The evolution of Chinese noodles has been described by Wei et al. (2017). The prototype of Chinese noodles is considered to be *Bing*, a form of flatbread, while knife-cut buckwheat noodles represent one of their earliest varieties. The first form of extruded noodles was *Hele* noodles, which were produced on a large scale. In addition, handmade noodles, which were typically made at home, evolved from processing techniques related to knife-cut buckwheat noodles and laomian. These noodles began to develop after wheat was introduced to the central plains of China, as well as the western and central Shaanxi plains (Hou 2020).

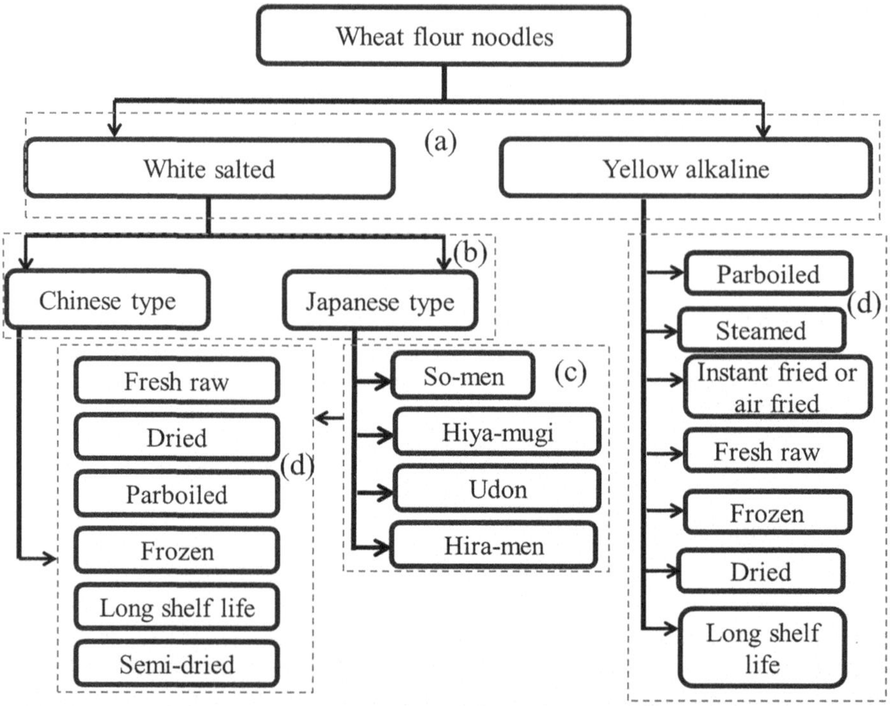

Fig. 2.1 Classification of wheat flour noodles. (**a**) Based on the absence or presence of alkaline salts; (**b**) According to the type of wheat flour; (**c**) Based on the size of noodle strands; and (**d**) According to the processing techniques

2.7 Rice Noodles and Other Starch Noodles

In China, rice noodles are referred to as *mifen*, where *"mi"* denotes rice and *"fen"* refers to starch noodles. These noodles are available in two main forms: *hefen* (flat strips) and *mixian* (thin threads). Based on their preparation methods, rice noodles are further classified into two categories: *zhafen* and *qiefen* (Fig. 2.2). "*Zha*" signifies "extruding into threads," resulting in noodles with a round cross-section, whereas "*Qie*" means "slicing broad strips from a thin, large sheet". A specific type, such as *changfen* or *juanfen*, is commonly used to wrap fillings. Most rice noodles are non-fermented (Song et al. 2024); however, *zhafen* is typically a fermented product. Fermented rice noodles, also known as sour *mifen*, are characterized by their slightly sour flavor.

It is believed that rice noodles originated during the Qin dynasty (259–210 B.C.) in China more than 2000 years ago and were introduced into neighbouring countries (Lu and Nip 2006). The diverse flavors and convenience of preparation have made rice noodles a popular choice in both restaurants and home kitchens in southern China (Phan 2020). Rice noodles are versatile and can be served as either a side dish

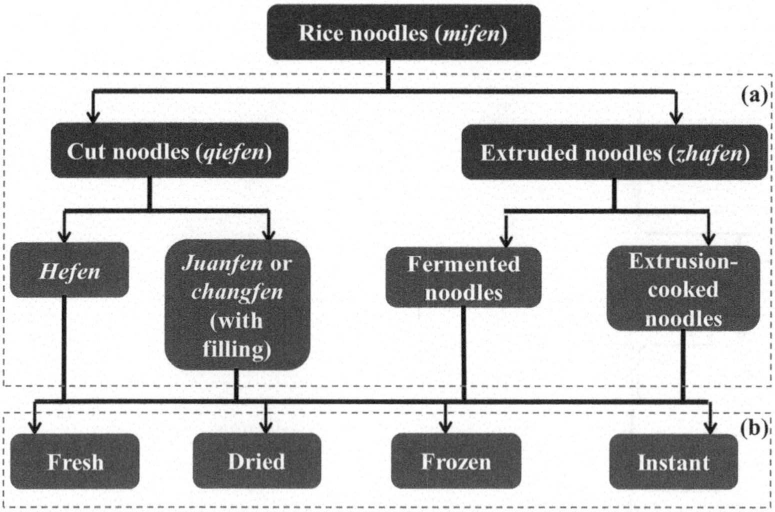

Fig. 2.2 Classification of rice noodles. (**a**) Based on noodle shaping methods; (**b**) Based on processing methods and moisture content

or a staple food. They are commonly prepared as soup noodles, combined with vegetables and meats, and cooked in broth, or they can be stir-fried as a flavorful alternative.

Traditionally, rice noodles were made using hand tools. However, modern rice noodle production is largely mechanized and utilizes automated production lines. The manufacturing process involves several key steps, including raw rice selection, cleaning, soaking, grinding, heating (by boiling or steaming), molding (through extrusion or cutting), cooling, and drying (Fig. 2.3). Table 2.3 shows the starch noodles from crops other than rice.

2.8 Gluten-Free Noodles

Gluten-free noodles are a nutrient-rich food option, particularly beneficial for individuals with gluten intolerance or celiac disease (Kaur et al. 2024). In Asia, a variety of raw materials such as konjac, cassava, corn, mung beans, sweet potatoes, potatoes, buckwheat, oats, and rice are commonly used to produce gluten-free noodles (Tong 2020). The network structure of gluten-free noodles largely relies on starch gelatinization, which binds the other ingredients together (Tong 2020). The rising demand for gluten-free products is attributed to the increasing prevalence of gluten sensitivity and celiac disease (Singla et al. 2024). Gluten, an essential protein in wheat, forms disulfide bonds and noncovalent interactions between glutenins and gliadins, creating an elastic network that provides structure and texture to doughs (Gasparre and Rosell 2023).

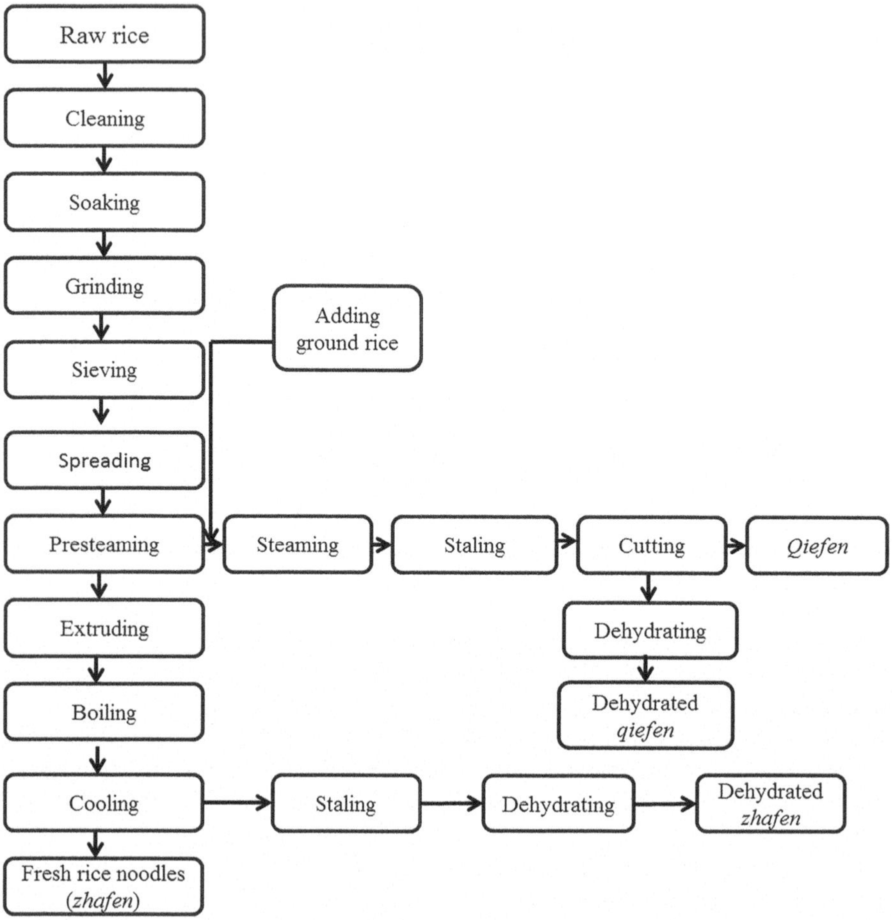

Fig. 2.3 Processing steps in the production of different types of rice noodles. (Source: Li et al. (2015))

Table 2.3 Starch noodles from crops other than rice

Crops	Starch noodles (English names)	Starch noodles (regional country names)
Yam	Devil's tongue noodles, Yam noodles	*Ito Konnyaku, sirataki, shirataki* (Japan)
Arrowroot	Arrowroot starch noodles	bôt mì tinh, Bôt hoành tinh, Bôt dong (Vietnam)
Cassava	Tapioca sticks	*Hu tieu bot loc* (Vietnam)
Sweet potato, potato	Sweet potato noodles	*Tang myun, Dang myeun* (Korea)
Canna	Canna Starch noodles	*Mién* dong, *mién* (Vietnam)
Buckwheat	Buckwheat noodles	Soba (Japan), *Naeng myon* (Korea)

(continued)

Table 2.3 (continued)

Crops	Starch noodles (English names)	Starch noodles (regional country names)
Corn	Corn starch noodles	*Pancit Luglug* (Philippines)
Mung bean, pea	Jelly noodles	*Harusame* (Japan)
	Crystal, glass noodles	*Bun tau* (Vietnam)
	Shining noodles	*Woon sen* (Thailand)
	Silver noodles	*Sotanghon* (Philippines)
	Translucent noodles	*Tung boon* (Indonesia)
	Vermicelli	*Soo hoon, Su boon* (Cantonese)
	Green bean thread noodles	*Fensi, Fen szu, Sai fun, Bai fun* (Chinese)

Source: Alden (2005); Lu and Collado (2010)

References

Ahmed Z, Chen J, Tufail T et al (2024) Fundamental opportunities and challenges of nutraceutical noodles enriched with agri-food by-products. Trends Food Sci Technol 143:104299

Alden L (2005) The Cook's Thesaurus. Available from http://www.foodsubs.com/NoodlesAsianOther.html

Crosbie GB, Ross AS, Moro T et al (1999) Starch and protein quality requirements of Japanese alkaline noodles (ramen). J Cereal Chem 76:328–334

Donadio R (2009) So you think you know pasta. New York Times. http://www.nytimes.com/2009/10/14/dining/14ency.html?_r=1. Accessed on 14 Oct 2009

Du J, Li Q, Obadi M et al (2023) Quality evaluation systems and methods of the whole making process of Asian noodles: a review. Food Rev Int 39:3830–3857

Fu BX (2008) Asian noodles: history, classification, raw materials, and processing. Food Res Int 41:888–902

Gao Y, Wang X, Shi Y (2022) The types, regional distribution, and consumption trend of Chinese traditional wheat-based foods. J Food Qual 2022:9986119, 10 pages

Gasparre N, Rosell CM (2023) Wheat gluten: a functional protein still challenging to replace in gluten-free cereal-based foods. Cereal Chem 100:243–255

Hou GQ (2001) Oriental noodles. Adv Food Nutr Res 43:141–193

Hou GG (2010) Asian noodles: science, technology, and processing. John Wiley & Sons, Inc, Hoboken, p 439

Hou GG (2020) Chapter 1 Introduction to Asian noodles. In: Hou GG (ed) Asian noodle manufacturing: ingredients, technology, and quality. Cereals & Grains Association: Woodhead Publishing, Duxford, pp 1–12

Hou G, Kruk M (1998) Asian noodle technology. AIB Technical Bulletin. Volume XX(12). American Institute of Baking, Manhattan, KS, Table II, p 2

Kaur S, Kumar K, Singh L et al (2024) Gluten-free grains: importance, processing and its effect on quality of gluten-free products. Crit Rev Food Sci Nutr 64:1988–2015

Li Y, Liang J-F, Yang M-Y et al (2015) Traditional chinese rice noodles: History, classification and processing methods. Cereal Foods World 60:123–127

Lu Z-H, Collado LS (2010) Rice and starch-based noodles. In: Hou GG (ed) Asian noodles: science, technology, and processing. John Wiley & Sons, Inc, Hoboken, pp 393–431

Lu S, Nip W-K (2006) Manufacture of Asian (Oriental) noodles. In: Hui YH (ed) Handbook of food science, technology, and engineering. Taylor & Francis, Boca Raton

Lu H, Yang X, Ye M et al (2005) Culinary archaeology: millet noodles in Late Neolithic China. Nature 437:967–968

Mahmud N, Islam J, Tahergorabi R (2023) Noodles. In: Shah MA, Sunooj KV, Mir SA (eds) Cereal-based food products. Springer, Cham

Meenu M, Padhan B, Zhou J et al (2023) A detailed review on quality parameters of functional noodles. Food Rev Int 39:6162–6198

Nagao S (1996) Processing technology of noodle products in Japan. In: Kruger JE, Matsuo RB, Dick JW (eds) Pasta and noodle technology. American Association of Cereal Chemists, St. Paul, pp 169–194

Obadi M, Xu B (2021) Review on the physicochemical properties, modifications, and applications of starches and its common modified forms used in noodle products. Food Hydrocoll 112:106286

Obadi M, Zhang J, He Z et al (2022a) A review of recent advances and techniques in the noodle mixing process. LWT 154:112680

Obadi M, Zhang J, Xu B (2022b) The role of inorganic salts in dough properties and noodle quality—a review. Food Res Int 157:111278

Obadi M, Guo Q, Sun J et al (2024) Recent developments in the application of physical processing techniques for controlling browning in fresh wet noodles: a review. J Cereal Sci 118:103951

Phan UTX (2020) Meals and snacks in Southeast and East Asia. In: Meiselman H (ed) Handbook of eating and drinking. Springer, Cham

Sadiku MNO, Chukwu UC, Sadiku JO (2023) Traditional Japanese food. Cent Asian J Lit Philos Cult 4:8–21

Singla D, Malik T, Singh A et al (2024) Advances in understanding wheat-related disorders: a comprehensive review on gluten-free products with emphasis on wheat allergy, celiac and non-celiac gluten sensitivity. Food Chem Adv 4:100627

Song Y, Lin J, Luo L et al (2024) Solid-state fermentation of Limosilactobacillus fermentum and lysine addition to improve the cooking and eating quality of high-fiber rice noodles. J Cereal Sci 118:103959

Tong L-T (2020) Chapter 7 – Gluten-free noodles. In: Hou GG (ed) Asian noodle manufacturing: ingredients, technology, and quality. Woodhead Publishing, Duxford, pp 125–149

Wei Y, Zhang Y, Liu R et al (2017) Origin and evolution of Chinese noodles. Cereal Foods World 62:44–51

Zhang M, Ma M, Yang T et al (2022) Dynamic distribution and transition of gluten proteins during noodle processing. Food Hydrocoll 123:107114

Chapter 3
Processing of Rice Noodles

Contents

Rice noodle is a gluten-free food and is one of the most popular noodles in Asia. Rice noodles are among the most widely consumed and produced rice-based staple foods worldwide, especially in Asian countries such as Korea, Japan, and China (Li et al. 2021). It is the most frequently consumed rice product after cooked rice grain in Asia (Du et al. 2023). The market size of rice noodles in the Europe and Asia Pacific regions is huge, and the expansion of markets in India and China is projected to increase sales worldwide. Rice noodle is made by mixing water and rice flour containing a high amylose content (>25%) (Wei et al. 2022). The rice flour slurry is steamed to obtain the fresh rice noodle sheet. Subsequently, the fresh noodle sheet is cut into strips. In general, fresh rice noodle exerts high moisture content and thus it is susceptible to being perishable. To prolong the shelf-life, the fresh rice noodle is dried to the final moisture content of 10–12%. In general, the drying process of rice noodles is categorized into three steps including final drying, aging, and pre-drying. During the pre-drying step, a fraction of water is removed, thus resulting in the moisture content of the noodle being about 34% (Sangpring et al. 2015). Subsequently, the noodle sheet is aged for 12–24 h at room temperature. Lastly, the noodle sheets are cut into strips and dried to the final moisture content at 10–12% (Sangpring et al. 2015). Rice noodles can be classified into fresh (>60% moisture), dry (<14% moisture), and semi-dried (14–60% moisture) varieties, depending on

B. L. Tan, M. E. Norhaizan, *Rice Noodles: Quality Improvement, Processing Techniques and Nutritional Aspects*, SpringerBriefs in Food, Health, and Nutrition,
https://doi.org/10.1007/978-3-031-98649-9_3

their moisture content levels (Li et al. 2015). Semi-dried rice noodles, which undergo moderate drying, maintain the flavor and texture qualities of fresh rice noodles while offering an extended shelf life similar to that of dry rice noodles (Zhu et al. 2023).

3.1 Starch: A Brief Description

Among the ingredients in rice noodles, starch is the largest component in frequently consumed fresh rice noodles. The physicochemical properties of starch are crucial in determining the quality of rice products (Rostamabadi et al. 2024). Amylose is usually chosen as a sensitive index to reflect the rice noodles quality (Zhu et al. 2024) because it plays a crucial role in forming the starch network (Tian et al. 2023a, b). Amylopectin contributes to nearly 70–99% of the total starch and plays a vital role in the physicochemical properties of rice starch and forming the crystal structure of starch granules (Shoukat et al. 2025). A study by Yi et al. (2021) evaluated the starch structure and its physicochemical properties from five rice cultivars ('NR', 'BX02', 'Zao 87', 'Zhongzao 35', and 'GM03') with different amylose contents in the formation of fresh rice noodles. The study showed the relative crystallinity of the starch samples did not affect the fresh rice noodles formation (Yi et al. 2021). However, the gel texture of starch played a crucial role in the fresh rice noodles formation. The study further revealed that fresh rice noodles made from NR rice cultivars demonstrate the excellent texture of fresh rice noodles. The morphology of fresh rice noodles from the GM03 cultivar was closest to NR but had lower cohesiveness, resilience, and springiness and higher hardness (Yi et al. 2021). The extracted starch from BX02, Zao87, and Zhongzao 35 rice cultivars, with lower fractions of amylose or higher swelling power and higher proportions of amylopectin, could not produce a consistent and dense fresh rice noodle structure (Yi et al. 2021).

3.2 The Form of Starch

Starch is the primary storage carbohydrate in plants, found as insoluble, semi-crystalline granules in storage tissues such as grains, tubers, and roots, and to a lesser extent in most vegetative tissues (de Souza Moretti et al. 2023). It consists of two polymers of D-glucose: amylose, an essentially unbranched $\alpha(1\rightarrow4)$-linked glucan, and amylopectin, a highly branched polymer with $\alpha(1\rightarrow4)$-linked glucose chains and $\alpha(1\rightarrow6)$ branching points. Native starch granules typically contain about 10% moisture. Amylose and amylopectin account for 98–99% of the dry weight of starch granules, with the remaining 1–2% comprising lipids, minerals, and phosphorus in the form of phosphate esters attached to glucose hydroxyl groups (Olawoye et al. 2023). Starch granules vary significantly in size (ranging from 1 to 100 μm in diameter), shape (polygonal, spherical, or lenticular), and molecular organization, including the content and structure of amylose and amylopectin, the branching architecture of amylopectin, and the degree of crystallinity (Ai and Jane 2024). Granules can occur individually or as compound clusters, and in grains like wheat, barley, rye, and triticale, they often display bimodal size distributions (Ledezma 2024). The diverse physical and molecular characteristics of starch granules significantly influence their functionality in various applications.

Although lipids constitute a minor component of starch by weight, they play a significant role in determining its properties. The lipid content and composition of starch granules vary across plant species (Sengupta et al. 2023). In isolated cereal starch granules, lipids are found both on the surface and within the granule. Surface lipids primarily consist of triglycerides, along with smaller amounts of free fatty acids, glycolipids, and phospholipids, which can be extracted using diethyl ether (Soman et al. 2024). Internal lipids in cereal starches are predominantly monoacyl lipids, typically extracted with hot aqueous alcohol. Both surface and internal lipids may exist in the free state, bound through ionic or hydrogen bonds to hydroxyl groups, or incorporated as amylose-lipid inclusion complexes (Ledezma 2024). The lipid content of native starch is closely linked to its amylose content, with higher amylose levels correlating to increased lipid content.

Amylose is a polymer with a molecular weight ranging from 10^5 to 10^6, corresponding to a degree of polymerization (DP) of 1000–10,000 glucose units. Its structure has a low level of branching, with less than 0.5% of glucose molecules linked by α(1→6) bonds. Typically, amylose consists of 3–11 chains, each containing 200–700 glucose residues. Due to its low branching, dissolved amylose tends to form insoluble semi-crystalline aggregates, influenced by the arrangement of its branches (Mhaske et al. 2024).

Amylopectin, by contrast, is a much larger polymer with a molecular weight of approximately 10^8 and a DP that may exceed one million. It constitutes 60–90% of most starches, although variations exist, such as high-amylose starches (as low as 30% amylopectin) and waxy starches (nearly 100% amylopectin). About 5% of glucose residues in amylopectin are linked by α(1→6) bonds, giving it a highly branched, tree-like structure. The branching pattern includes A-chains (unsubstituted), B-chains (substituted by other chains), and a single C-chain (carrying the reducing glucose). This complex architecture varies among starches based on the placement and length of branches. Glycogen, the storage polysaccharide in animals, shares the same chemical composition as amylopectin but has a higher degree of branching, resulting in a more compact, globular shape (Da Poian and Castanho 2021). Amylopectin's tree-like structure allows more glucose to be stored at branch ends compared to glycogen's denser, globular configuration. The diversity in amylose and amylopectin is attributed to the intricate biosynthesis of starch. This process involves several key enzymes including ADP-glucose pyrophosphorylases, which produce ADP-glucose from hexose-phosphates, starch synthases (both soluble and granule-bound), which form α(1→4) linkages, starch branching enzymes, catalyze the formation of α(1→6) branch linkages, and starch debranching enzymes, which trim excess branches to facilitate amylopectin crystallization. These enzymes occur in multiple forms and are influenced by genetic and environmental factors. Their activities can vary depending on their temporal and spatial expression, as well as substrate specificities. The final starch composition reflects the genetic diversity of these enzymes and the environmental conditions affecting their activity (Rhazi et al. 2021).

3.3 Starch Granules

The crystallinity of native starch granules varies significantly, ranging from approximately 15% in high-amylose starches to 45–50% in waxy starches (Obadi et al. 2023). These granules exhibit a hierarchical structure, which can be visualized using light and electron microscopy. They consist of multiple concentric layers, known as growth rings, which radiate outward from the hilum (the growth center) to the surface of the granule. The growth rings, typically 120–400 nm thick, are thought to represent daily fluctuations in starch deposition within storage tissues (Huang et al. 2022). Within the growth rings, there are alternating crystalline and amorphous regions, corresponding to areas of higher and lower density, respectively. The crystalline regions exhibit a lamellar structure, consisting of alternating crystalline and amorphous layers with a periodicity of 9–11 nm (Sun et al. 2021). The crystalline layers are primarily formed by amylopectin chains arranged in a crystalline lattice, while the amorphous layers contain amylopectin branching points along with disordered amylose and amylopectin molecules. Longer amylopectin chains extend from the crystalline into the amorphous regions of the lamellae (Zhong et al. 2021). Near the granule surface, amylopectin molecules may differ structurally from those closer to the hilum. The ratio of long to short branch chains influences the overall shape of amylopectin and its packing within the granules (Ai and Jane 2024). Genetic and environmental factors that impact amylopectin biosynthesis play a critical role in shaping its molecular architecture, which in turn affects the morphology of starch granules (Shi et al. 2023).

Amylose is primarily located in the low-density regions of the growth rings. However, amylose molecules are also thought to be dispersed within the crystalline layers of amylopectin, disrupting its crystal packing (Wang et al. 2024a, b). In high-amylose starches, the amylose helices may contribute to the granule's overall crystallinity (Tian et al. 2024). The arrangement of amylose and amylopectin within the growth rings, as well as the lamellar architecture of the crystalline layers, remains incompletely understood. Similarly, the distribution and role of water within granules, which acts as a plasticizer and influences crystallinity required further investigation (Li et al. 2023). Granules typically contain approximately 10% water, although its distribution is likely non-uniform.

Amylopectin chains containing more than 10 glucose units form double helices, which are arranged into either A- or B-type crystalline forms, identifiable by characteristic X-ray diffraction (XRD) patterns (Ukkunda et al. 2024). Although the double helices in both A- and B-type crystalline forms share a similar structure, the packing of the helices differs. The A-type crystals have a more compact arrangement, while the B-type crystals feature a more open structure with a hydrated core (Koekuyt and Marangoni 2024). A-type crystalline patterns are typically associated with amylopectin molecules that have shorter chains (Nakamura et al. 2023). These patterns are common in cereal starches, whereas tuber and amylose-rich starches generally exhibit B-type patterns. An intermediate C-type pattern is observed in legume, root, and some fruit and stem starches.

3.4 Processing of Rice Noodles

There are several key steps involved in raw rice processing into rice noodles including selecting raw rice, followed by cleaning, soaking, grinding, heating (boiling or steaming), molding (cutting or extruding), and cooling (drying) (Table 3.1).

Several steps are involved in the red rice processing into red rice noodles including mixing, steaming, gelatinizing tapioca starch, combining all constituent materials, sheeting the red rice mixtures into noodle form, additional steaming of the noodles, and drying. The first step in the noodle processing is mixing, in which the *Vigna radiata* sprout flour and red rice sprout flour are added to water. The mixing of these ingredients generated a uniform distribution of materials and gave hydration to the flour. The steaming process takes nearly 40–45 min, in which the starch of brown rice partially gelatinizes during the steaming process (Nugraheni et al. 2022). Subsequently, the dough was kneaded to distribute the partially gelatinized starch. Tapioca flour is used to enhance the texture of noodles and serves as an adhesive, and thereby gelatinization is performed in tapioca flour (Obadi and Xu 2021). The manufacturing process is initiated with the creation of tapioca flour that is gelatinized. This process is accomplished by mixing water with tapioca flour and heating it until it turns clear (gelatinized).

In general, the rice noodle manufacturing process generates significant amounts of starchy waste. Raw rice flour wastewater is produced during the washing, soaking, and milling stages, while large quantities of gelatinized flour are discarded daily after the steaming of rice flour slurry (Kanpiengjai et al. 2014). Due to this substantial generation of starchy waste, implementing proper waste treatment strategies is essential. Bioconversion of bio-waste into higher value-added products, such as biofuel and organic acids, has been extensively studied and promoted (Pandey et al. 2000; Ohkouchi and Inoue 2007). With the growing global demand for lactic acid used in the food, textile, leather, and electronic industries, hence the conversion of starchy waste into lactic acid is of considerable interest. Direct fermentation of starch into lactic acid using amylolytic lactic acid bacteria (ALAB) has been proposed as an alternative strategy. This approach eliminates the need for additional starch liquefaction and saccharification steps, as ALAB can directly convert starch into lactic acid. The strategy has been evaluated for its potential, particularly in addressing the high demand for amylase enzymes (Kanpiengjai et al. 2014). Therefore, directly converting starchy wastewater and gelatinized starchy waste into lactic acid represents a value-adding solution with significant potential for lactic acid applications across various industries.

Table 3.1 Rice noodles processing

Stage	Description	References
Selection of raw rice	Rice noodles are made with milled rice grain, in which components such as ash, low molecular weight sugars, lipids, proteins, and starches may affect the consistency of noodles. Traditionally, the *indica* variety is used in rice noodle formulations. A mixture of early- and late-harvest *indica* rice is usually used to produce noodles	Ding et al. (2004)
Cleaning raw rice	Removing foreign materials and dust	Li et al. (2015)
Soaking	Soaking for a short time allows water to penetrate into the rice kernel, thus leading to high moisture content and a less rigid structure without initiating fermentation	Cen (2007)
	Soaking rice for a long time is essential for producing naturally fermented rice noodles. Fermentation is facilitated by microbes, for instance, yeasts and lactic acid bacteria	Liu (2007)
Sieving and grinding	Once the soaked rice achieves the appropriate moisture content, it is ground and sieved (after dry grinding)	Li et al. (2015)
	Dry grinding—A hammer mill is commonly used for dry grinding soaked rice. When ground to a proper size, the rice particles can pass through 60 or 80 sieves. Typically, rice soaked for a short time is suitable for dry grinding; whereas dry-grinding rice with higher moisture content (>26–28%) may cause rice agglomeration or plugging of the mesh sieve	Pan (2007)
	Wet grinding—Rice soaked for a long time (containing about 45%) is suitable for wet grinding. Wet grinding produces a rice slurry with a smooth consistency	Li et al. (2015)
Molding	Shaping of ground rice. Molding can be classified into two techniques, namely cutting and extrusion	Li et al. (2015)
	Extruding rice noodles (*zhafen*)—Extruding is a crucial molding technique that produces rice noodles with a rounded shape. Rice slurry is transported by a band carrier in sheets (about 5 mm thick). The sheet is pre-steamed for 4–5 min at 90 °C. Lastly, the steamed sheets are extruded into rice noodles and boiled to ensure they have completely gelatinized	Li et al. (2015)
	Cutting rice noodles (*Qiefen*)—Before cutting, ground rice is undergone two heating processes. Initially, a portion of the ground rice is pre-steamed, and the remainder of the ground rice is added to make a thick rice slurry and formed 1–3 mm thick sheets. The sheets are fully steamed, and band carriers transport the steamed rice sheets to be cut into rectangular noodles	Li et al. (2015)
Heating (Boiling or streaming)	Gelatinize the rice starch to its α-state	Pan (2007), Pan and Deng (2002)
Cooling	Placing boiled or steamed rice noodles in cool water (0–10 °C) to reduce temperature to 24–26 °C and form a strong gel. The cooling procedure facilitates the noodle separation and thus creates a slippery, smooth noodle texture	Li et al. (2015)

(continued)

Table 3.1 (continued)

Stage	Description	References
Staling	Steamed, gelatinized rice noodles are placed in a special room for 12–24 h to convert starches to β-state. This process increased flexibility, improved structure stability, and non-sticky noodles	Li et al. (2015)
Dehydrating	Rice noodles may be dehydrated to a low moisture content to prolong shelf life. Drying reduces the moisture content in rice noodles from 40–65% to 15–25%. There are two types of equipment widely used, namely drying room and continuous drying cabinet	Pan and Deng (2002)

3.5 Industrial Production of Rice Noodles

Rice noodles are a significant staple food in Southeast Asian countries and are typically made from Indica rice. The production methods can be categorized into fermented and unfermented processes. The network structure of rice noodles largely depends on the starch content, composition, and physicochemical properties (Yi et al. 2020a, b). Several studies have demonstrated that adding zein and gums enhances the texture and cooking qualities of rice noodles (Zhang et al. 2023). In fermented food production, microorganisms play a crucial role in shaping the texture, flavor, and safety of the final product (Abbaspour 2024).

Flavor is a key determinant of the quality of rice products. The volatile components of these products are influenced by factors such as raw materials, milling degree, and fermentation methods (Hu et al. 2020). The bacterial composition of fermented rice slurry significantly impacts the volatile components of fermented rice noodles (Wang et al. 2022). Numerous studies have shown that natural fermentation involves complex microbial and enzymatic processes, which contribute to the distinctive flavor of rice noodles (Yi et al. 2020a, b; Sun et al. 2025). However, the relationship between the volatile components of fermented rice noodles and the composition of microbial communities remains unclear. In most rice noodle factories, rice grains are typically soaked in large parallelepiped steel tanks covered with a thin layer of water (8–15 cm) and fermented at ambient temperature for 2–3 days. This prolonged soaking period often results in harmful microbial growth and contributes to environmental pollution (Roy et al. 2011).

In addition, natural fermentation is highly sensitive to environmental changes, leading to inconsistent product properties and significant variations in rice noodle characteristics across different regions. As a result, natural fermentation fails to meet consumer standards and industrial demands for consistent and high-quality products (Steensels and Verstrepen 2014). Traditionally, rice grains are fermented for several days without the use of a starter culture before being processed into naturally fermented rice noodles (Lu et al. 2008). The fermentation process alters the concentrations of free amino acids and organic acids in rice slurry, as well as the proximate composition and pasting properties of rice flour (Ohishi et al. 2007; Guimaraes et al. 2017; Zhu et al. 2016). Free amino acids produced during lactic acid bacteria fermentation significantly enhance the flavor of fermented foods (Irigoyen et al. 2007). Lactic acid is the primary organic acid generated during fermentation, lowering the pH and inhibiting the growth of harmful microorganisms (Guimaraes et al. 2017; Aline et al. 2018). Furthermore, the pasting viscosities of rice flour decrease following lactic acid bacteria fermentation (Yang and Tao 2008). Fermented rice noodles exhibit improved texture, enhanced eating qualities, and a distinctive flavor compared to unfermented rice noodles (Lu et al. 2003; Keatkrai and Jirapakkul 2010). However, the natural fermentation process requires rice to be soaked for 2–3 days, which elevates the risk of harmful microbial contamination and degrades the quality of the rice noodles (Charles et al. 2007).

To mitigate the adverse effects of prolonged soaking times, the dry milling method has been proposed as a viable alternative for breaking rice granules (Fang et al. 2023). However, this approach increases the content of damaged starch and compromises the network structure of rice noodles compared to the wet milling method (Ronie and Hasmadi 2022). Consequently, dry-milled flour is not suitable for rice noodle processing.

The previous research demonstrated that semidry milling at 30% moisture effectively preserves the physicochemical properties of rice flour, reduces damaged starch content, and enhances rice noodle quality, yielding results comparable to those of wet milling (Yu et al. 2024). To further reduce soaking times, minimize bacterial contamination, and improve rice noodle safety, additional studies were conducted. The findings revealed that soaking times for Indica rice could be reduced to 10 min by inducing grain cracks through drying at 60 °C (Wang et al. 2024a, b). Moreover, hot air treatment was shown to shorten soaking times to 8 minutes, enabling rice grains to achieve maximum water content suitable for semidry milling (Huang et al. 2024). These methods not only reduced soaking time but also produced rice noodles with textures comparable to those obtained via wet milling.

In addition to physical treatments, pure bacterial fermentation was found to accelerate the fermentation process. Lactic acid bacteria and yeast are the dominant microorganisms in naturally fermented rice slurry (He et al. 2022), with *Lactobacillus* identified as an ideal strain for rice noodle fermentation (Geng et al. 2020). Using a starter culture is an effective strategy to accelerate the fermentation process in producing fermented rice noodles. *Lactobacillus plantarum*, the predominant bacterium in naturally fermented rice slurry, is considered an ideal strain for rice noodle fermentation due to its ability to create a pronounced fermented rice aroma (Yi et al. 2017, 2019). Our previous study found that pure bacterial fermentation reduced the fermentation time from 54 h, as seen in natural fermentation, to just 18 h, while also minimizing the risk of contamination by undesirable bacteria (Geng et al. 2019).

Starter cultures are commonly applied in the production of fermented foods like yogurt, bread, and steamed bread, where they not only speed up fermentation but also enhance product quality and safety (Kimaryo et al. 2000). Research has shown that *Lactobacillus plantarum* inoculum can effectively shorten fermentation times, suppress harmful microorganisms, and elevate levels of lactic acid and aminopeptidase activity in fermented foods (Sawatari et al. 2005; Avila et al. 2005). Despite these advancements, few studies have investigated the effects of pure starter cultures on the quality of fermented rice noodles. A study by Geng et al. (2020) demonstrated that the natural inoculum (10%) and pure inoculum (10%) fermentation methods significantly reduced the fermentation time to 20 and 10 h, respectively. The data showed that both natural and pure inoculums accelerated the fermentation rate of rice noodles. This acceleration may be attributed to the higher relative abundance of organic acid-producing bacteria in the inoculated rice slurries, which facilitated faster fermentation.

The concentrations of free amino acids and organic acids in rice slurry were observed to change during the fermentation process (Zhu et al. 2016; Guimaraes et al. 2017). Following fermentation, the levels of Ser, Glu, Leu, Tyr, Phe, and Arg increased in both naturally fermented and natural inoculum-fermented rice slurries. In contrast, in pure and pure inoculum-fermented rice slurries, the concentrations of all free amino acids increased, except for Pro (Geng et al. 2020). The free amino acid profile of the pure inoculum-fermented rice slurry closely resembled that of the pure fermented rice slurry. Notably, the concentrations of Asp, Glu, Leu, Tyr, Phe, Lys, and Arg were higher in the natural inoculum-fermented rice slurry compared to the naturally fermented slurry. Previous studies have demonstrated that *Lactobacillus plantarum* enhances the total free amino acid content in fermented products such as cheese, wheat dough, and meat (Irigoyen et al. 2007; Valerio et al. 2017; Fadda et al. 2008). It is likely that dominant bacteria, such as *L. plantarum*, in the inoculum-fermented rice slurries contributed to increased free amino acid production. Free amino acids play a key role in the flavor of rice noodles (Irigoyen et al. 2007), with branched-chain amino acids (Val, Ile, Leu), aromatic amino acids (Tyr, Phe), and Met serving as precursors for key aroma compounds in cheese (Yvon and Rijnen 2001). The concentrations of Val, Ile, Leu, Tyr, Phe, and Met were higher in the pure and pure inoculum-fermented rice slurries compared to the naturally fermented slurry. Thus, these amino acids may be critical in enhancing the flavor of rice noodles produced from pure and pure inoculum-fermented rice slurries. In addition, the lactic acid content increased by 98% and 20% in the natural inoculum and pure inoculum-fermented rice slurries, respectively, compared to their non-inoculated counterparts (Geng et al. 2020). Consistent with previous findings, *Lactobacillus plantarum* appears to enhance lactic acid production during rice fermentation (Sankar et al. 2018). The higher lactic acid content in inoculum-fermented slurries likely contributed to lowering the pH, thereby inhibiting the growth of undesirable bacteria (Guimaraes et al. 2017).

Research evidence has demonstrated that the total starch content in fermented rice flour was higher than in unfermented flour (Geng et al. 2020), likely due to the purification of rice starch during fermentation, which reduces other components (Lu et al. 2003). The damaged starch content in naturally and natural inoculum-fermented rice flours decreased post-fermentation (Geng et al. 2020), suggesting that natural fermentation may alter the rice grain structure and protect starch from damage during the milling process. In contrast, the amylose content in pure and pure inoculum-fermented rice flours decreased after fermentation (Geng et al. 2020). This reduction is consistent with previous findings, which suggest that amylose molecules may be broken down due to increased acidity, enzymatic activity, or consumption by *Lactobacillus plantarum* (Wan et al. 2011; Geng et al. 2019). The protein content was higher in pure and pure inoculum-fermented rice flours compared to natural and natural inoculum-fermented flours (Geng et al. 2020). This result indicates that proteins may dissolve in the soaking water and be lost during washing. Furthermore, it aligns with prior studies showing that Clostridium, a key bacterium in natural fermentation, can produce extracellular enzymes that degrade proteins into fermentable components (Geng et al. 2019; Tracy et al. 2012). In addition, the lipid content of fermented rice flour decreased significantly after fermentation (Geng et al. 2020). This reduction is likely due to the breakdown of lipids into fatty acids, which are subsequently converted into volatile compounds (Lu et al. 2008). The pasting properties of rice flour play a critical role in determining the texture, cooking characteristics, and eating quality of rice noodles (Tong et al. 2014). Natural fermentation alters the pasting properties of rice flour, resulting in improved eating quality of the fermented rice noodles (Lu et al. 2005). Compared to naturally and purely fermented rice noodles, the natural inoculum and pure inoculum-fermented rice noodles exhibited a slightly higher water absorption rate, along with slightly lower cooking loss and water turbidity. These findings suggest that the use of inocula may improve the texture of fermented rice noodles (Yi et al. 2017).

Fresh rice noodles, with their high moisture content (>50 g/100 g), are highly susceptible to microbial spoilage. To address this issue and extend their shelf-life, methods such as the use of preservatives and thermal sterilization (95 °C for 30 min) have been employed (Tantala et al. 2022; Gao et al. 2024). More commonly, fresh rice noodles are dehydrated to produce dried rice noodles, which not only have an extended shelf-life but are also more convenient for distribution and transportation (Gao et al. 2023). Dried rice noodles are starch-based gels, where the quality primarily depends on starch properties. Studies have demonstrated that hydrothermal treatments offer a more effective approach to modifying the properties of rice noodles compared to the use of additives (Kasunmala et al. 2020; Gao et al. 2023) or chemically modified starches (Jia et al. 2022). Research evidence has shown that noodles made from heat-moisture-treated potato starch exhibited a lower breakage rate (2.22–8.89%) and cooking loss (4.08–5.89%) compared to noodles prepared with native starch (Yang et al. 2022). A recent study by Liu et al. (2024) investigated the impact of directly incorporating heat-moisture treatment (110 °C, 2 h) into the drying process on the properties of dried rice noodles. The study showed that the dried rice noodles demonstrated enhanced structural organization and improved thermal resistance. Dried rice noodles treated with heat-moisture treatment showed limited starch leaching during rehydration and a decrease in cooking loss. Rice noodles subjected to heat-moisture treatment at 20% moisture content exhibited a substantially higher cooking yield, nearly threefold increase in extensibility, and 1.68 times greater chewiness compared to the control (Liu et al. 2024).

Brown rice noodles are developed by combining the nutritional benefits of brown rice with the appealing taste of semi-dried noodles. However, similar to other brown rice-based products, the inclusion of rice bran in brown rice noodles can lead to a coarse texture, which may affect consumer acceptance. Additionally, brown rice noodles produced through direct brown rice milling are prone to degradation caused by lipid oxidation during storage (Zeng et al. 2023).

Several methods can be used to improve the quality of brown rice, such as pre-soaking high hydrostatic pressure, enzymatic treatment (Yu et al. 2015), milling (Kim and Lee 2013), fermentation (Xu et al. 2021), and germination (Chao et al. 2022). Fermentation, a traditional technique in food processing and preservation, not only enhances the nutritional and sensory qualities of food products but also extends their shelf life. This method is crucial in modern food production (Xu et al. 2021). Currently, most rice noodle production involves natural liquid-state fermentation of rice grains (LFRG), which is susceptible to bacterial contamination and off-putting odors during fermentation. In addition, this process increases cleaning costs, produces significant wastewater, and poses food safety risks. These challenges can be addressed by using specific microorganisms for controlled fermentation, which improves the consistency and safety of the process. The previous study on liquid-state fermentation showed that controlled fermentation can improve the processing quality of rice flour, enhancing the flavor and texture of fermented rice noodles while inhibiting the growth of harmful microorganisms (Li et al. 2021). In contrast, solid-state fermentation provides advantages, for instance, lower energy consumption, reduced wastewater production, and more sustainable practices (Zhang et al. 2022).

Limosilactobacillus fermentium L6 is the dominant strain in the fermentation broth of Changde Jinjian rice noodles, a traditional fermented rice noodle variety used to ferment brown rice flour under solid-state conditions. The data showed that fermentation with *Limosilactobacillus fermentium* L6 significantly improved the texture and cooking qualities of semi-dry brown rice noodles, addressing the challenge of their limited shelf life (Xiong et al. 2023). The data further demonstrated that *Limosilactobacillus fermentium* L6 improved the cooking and sensory properties of high-fiber rice noodles by incorporating solid-state fermentation with *Limosilactobacillus fermentum* and supplementing with lysine (Song et al. 2024). Weiling et al. (2024) found that the quality of brown rice noodles can be significantly improved by applying these three fermentation techniques, namely solid-state fermentation of rice grains, liquid-state fermentation of rice grains, and solid-state fermentation of rice flour. Intriguingly, the study showed that brown rice noodles processed through solid-state fermentation with rice grains showed a superior gel structure, increased luminosity (L*), and improved cooking and sensory attributes, suggesting that substantial potential for optimizing the production process of fermented brown rice noodles.

In general, the rehydration of dried rice noodles is a slow process, primarily influenced by the density of the noodle structure. Rehydration time is a key quality indicator that consumers pay close attention to when evaluating dried rice noodles. Various strategies have been developed to accelerate the rehydration process of dried rice noodles, including improving hydrophilicity (Horstmann et al. 2018), and increasing the degree of starch gelatinization (Thammathongchat et al. 2024; Xue et al. 2008). Among these, creating a porous structure is the most frequently reported method. This can be achieved through drying technologies, the addition of soluble additives, or enzymatic treatments. Drying techniques induce pore formation, though the pore characteristics vary significantly depending on the method. Vacuum freeze-drying and deep-frying are commonly used, as they rapidly vaporize internal water, creating numerous small pores. However, vacuum freeze-drying is cost-intensive, and deep-frying raises health concerns, limiting their widespread application (Taşeri et al. 2018). Incorporating additives like soybean oleogel and maltodextrin has been shown to reduce noodle compactness by forming pores through dissolution and leaching (Jang et al. 2016; Oh and Lee 2020). Enzymatic methods involving hydrolytic amylase and glycosyltransferase effectively loosen the noodle structure (Zhang et al. 2024). The use of branching enzymes from glycosyltransferases showed promise in regulating the compactness of starchy products (Tian et al. 2023a, b). Notably, a porous structure can reduce rehydration time by decreasing the time required for water to penetrate the noodle core, as well as the amount of water needed. Magnetic resonance imaging (MRI) revealed that although larger pores allow for greater water absorption, the time for water to reach the center of the noodle is longer compared to medium-sized pores. This is because water must first fill the larger peripheral pores before migrating inward. Small-angle X-ray scattering (SAXS) analysis showed that flexible starch molecular chains reduce the water required to achieve maximum tensile strain, thereby shortening the rehydration time (Zhang et al. 2024).

References

Abbaspour N (2024) Fermentation's pivotal role in shaping the future of plant-based foods: an integrative review of fermentation processes and their impact on sensory and health benefits. Appl Food Res 4:100468

Ai Y, Jane J-L (2024) Chapter 3: Understanding starch structure and functionality. In: Nilsson L (ed) Starch in food: Strcture, function, and applications, 3rd edn, pp 55–77

Aline GTM, Cíntia LR, Dias DR et al (2018) Combination of probiotic yeast and lactic acid bacteria as starter culture to produce maize-based beverages. Food Res Int 111:187–197

Avila M, Garde S, Medina M et al (2005) Effect of milk inoculation with bacteriocin-producing lactic acid bacteria on a *Lactobacillus helveticus* adjunct cheese culture. J Food Protect 68:1026–1033

Cen JJ (2007) The prospect of rice noodles industry, its technology and equipment (In Chinese). Sci Technol Cereals Oils Foods 15:4

Chao S, Mitchell J, Prakash S et al (2022) Effect of early harvest and variety difference on grain yield and pasting properties of Brown rice. Crops 2:23–39

Charles AL, Huang TC, Lai PY et al (2007) Study of wheat flour-cassava starch composite mix and the function of cassava mucilage in Chinese noodles. Food Hydrocoll 21:368–378

Da Poian AT, Castanho MARB (2021) The families of biological molecules. In: Integrative human biochemistry. Springer, Cham, pp 89–185

de Souza Moretti MM, Costa MS, Bai Y et al (2023) Chapter 9: Structure of starch, focusing on those from underground plant organs. In: Cereda MP, Vilpoux OF (eds) Starchy crops morphology, extraction, properties and applications. Vol 1: underground starchy crops of South American origin: production, processing, utilization and economic perspectives, pp 217–244

Ding WP, Wang YH, Xia WS (2004) Determination on choosing standard of raw material in manufacturing rice noodles (In Chinese). Food Sci Technol 10:24

Du J, Li Q, Obadi M et al (2023) Quality evaluation systems and methods of the whole making process of Asian noodles: a review. Food Rev Int 39:3830–3857
Fadda S, Chambon C, Champomier-Verges MC et al (2008) *Lactobacillus* role during conditioning of refrigerated and vacuum-packaged Argentinean meat. Meat Sci 79:603–610
Fang S, Chen M, Xu F et al (2023) The possibility of replacing wet-milling with dry-milling in the production of waxy rice flour for the application in waxy rice ball. Foods 12:280
Gao L, Xu Z, Zhang R et al (2023) Effects of erythritol on rheological properties of rice flour and structural characteristics of extruded dried rice noodles with rapid rehydration behaviors. Food Hydrocoll 144:109007
Gao L, Guan M, Qin Y et al (2024) Utilization of heat-induced curdlan gel to improve the cooking qualities of thermally sterilized fresh rice noodles. Int J Biol Macromol 262:129693
Geng DH, Liang T, Yang M et al (2019) Effects of *Lactobacillus* combined with semidry flour milling on the quality and flavor of fermented rice noodles. Food Res Int 126:108612
Geng D-H, Liu L, Zhou S et al (2020) Effects of *Lactobacillus plantarum* inoculum on the fermentation rate and rice noodle quality. J Oleo Sci 69:1031–1041
Guimaraes A, Santiago A, Teixeira JA et al (2017) Anti-aflatoxigenic effect of organic acids produced by, *Lactobacillus plantarum*. Int Food Microbiol 264:31–38
He L, Chen Y, Zhang H et al (2022) Isolation and identification of Lactobacillus and yeast species and their effect on the quality of fermented rice cakes. Innov Food Sci Emerging Technol 77:102984
Horstmann SW, Axel C, Arendt EK (2018) Water absorption as a prediction tool for the application of hydrocolloids in potato starch-based bread. Food Hydrocoll 81:129–138
Hu X, Lu L, Guo Z et al (2020) Volatile compounds, affecting factors and evaluation methods for rice aroma: a review. Trends Food Sci Technol 97:136–146
Huang J, Wang Z, Fan L et al (2022) A review of wheat starch analyses: methods, techniques, structure and function. Int J Biol Macromol 203:130–142
Huang T, Ouyang D, Sang S et al (2024) Effect of water content in semidry grinding on the quality of glutinous rice flour. Foods 13:3216
Irigoyen A, Ortigosa M, Juansaras I et al (2007) Influence of an adjunct culture of *Lactobacillus* on the free amino acids and volatile compounds in a Roncal-type ewe's-milk cheese. Food Chem 100:71–80
Jang A, Kim JY, Lee S (2016) Rheological, thermal conductivity, and microscopic studies on porous-structured noodles for shortened cooking time. LWT 74:1–6
Jia Y, Zhang Z, Li M et al (2022) The effect of hydroxypropyl starch on the improvement of mechanical and cooking properties of rice noodles. Food Res Int 162:111922
Kanpiengjai A, Lumyong S, Pathom-aree W et al (2014) Starchy effluent from rice noodle manufacturing process as feasible substrate for direct lactic acid production by *Lactobacillus plantarum* S21. J Korean Soc Appl Biol Chem 57:217–220
Kasunmala IGG, Navaratne SB, Wickramasinghe I (2020) Effect of process modifications and binding materials on textural properties of rice noodles. Int J Gastron Food Sci 21:100217
Keatkrai J, Jirapakkul W (2010) Volatile profile of khanom jeen, Thai fermented rice noodles, and the changes during the fermentation process. ScienceAsia 36:46–51
Kim H, Lee H (2013) Effects of eating quality on milled rice produced from brown rice with different milling conditions. J Korean Soc Appl Biol Chem 56:621–629
Kimaryo VM, Massawe GA, Olasupo NA et al (2000) The use of a starter culture in the fermentation of cassava for the production of "kivunde", a traditional Tanzanian food product. Int J Food Microbiol 56:179–190
Koekuyt HA, Marangoni AG (2024) The solid state and nanostructure of starch: effects on starch functionality. Crit Rev Food Sci Nutr. https://doi.org/10.1080/10408398.2024.2388279
Ledezma CCQ (2024) Chapter 6: Starch interactions: implications for food quality and innovation. In: Nilsson L (ed) Starch in food: structure, function, and applications, 3rd edn, pp 119–144
Li Y, Liang J-F, Yang M-Y et al (2015) Traditional chinese rice noodles: history, classification, and processing methods. Cereal Foods World 60:123–127
Li C, Chen Y, Gu D et al (2021) A systematic review of rice noodles: raw material, processing method and quality improvement. Trends Food Sci Technol 107:389–400

Li M, Daygon VD, Solah V et al (2023) Starch granule size: does it matter? Crit Rev Food Sci Nutr 63:3683–3703

Liu XX (2007) Reconstruction of rice noodle industry and technical innovation in China (In Chinese). Cereal Feed Ind 14:1

Liu Y, Luo Q, Chen J et al (2024) Physicochemical quality improvement of dried rice noodles by direct heat-moisture treatment during the drying process. J Cereal Sci 118:103986

Lu ZH, Li LT, Cao W et al (2003) Influence of natural fermentation on physico-chemical characteristics of rice noodles. Int J Food Sci Technol 38:505–510

Lu ZH, Li LT, Min WH et al (2005) The effects of natural fermentation on the physical properties of rice flour and the rheological characteristics of rice noodles. Int J Food Sci Technol 40:985–992

Lu ZH, Cao W, Peng HH et al (2008) Effect of fermentation metabolites on rheological and sensory properties of fermented rice noodles. J Sci Food Agric 88:2134–2141

Mhaske P, Majzoobi M, Farahnaky A (2024) Starch, modified starch, and extruded foods. In: Rosenthal A, Chen J (eds) Food texturology: measurement and perception of food textural properties. Springer, Cham, pp 393–420

Nakamura Y, Yashiro K, Matsuba G et al (2023) The relationship between the fine structure of amylopectin and the type of crystalline allomorph of starch granules in rice endosperm. Cereal Chem 100:721–733

Nugraheni M, Windarwati, Palupi S (2022) Gluten-free noodles made based on germinated organic red rice: chemical composition, bioactive compounds, antioxidant activity and sensory evaluation. Food Res 6:354–363

Obadi M, Xu B (2021) Review on the physicochemical properties, modifications, and applications of starches and its common modified forms used in noodle products. Food Hydrocoll 112:106286

Obadi M, Qi Y, Xu B et al (2023) High-amylose maize starch: structure, properties, modifications and industrial applications. Carbohydr Polym 299:120185

Oh I, Lee S (2020) Rheological, microstructural, and tomographical studies on the rehydration improvement of hot air-dried noodles with oleogel. J Food Eng 268:109750

Ohishi K, Kasai M, Shimada A et al (2007) Effects of acetic acid on the rice gelatinization and pasting properties of rice starch during cooking. Food Res Int 40:224–231

Ohkouchi Y, Inoue Y (2007) Impact of chemical components of organic wastes on (+)-lactic acid production. Bioresour Technol 98:546–553

Olawoye B, Jolayemi OS, Akinyemi TY et al (2023) Modification of starch. In: Sharanagat VS, Saxena DC, Kumar K et al (eds) Starch: advances in modifications, technologies and applications. Springer, Cham, pp 11–54

Pan SP (2007) The technology of fermented instant rice noodles (In Chinese). Jiangxi Food Ind 2:42

Pan SP, Deng DW (2002) High quality straight strip rice noodle making in Jiangxi (In Chinese). Jiangxi Food Ind 1:11

Pandey A, Soccol CR, Nigam P et al (2000) Biotechnological potential of agro-industrial residues. II: cassava bagasse. Bioresour Technol 74:81–87

Rhazi L, Méléard B, Daaloul O et al (2021) Genetic and environmental variation in starch content, starch granule distribution and starch polymer molecular characteristics of French bread wheat. Foods 10:205

Ronie ME, Hasmadi M (2022) Factors affecting the properties of rice flour: a review. Food Res 6:1–12

Rostamabadi H, Yildirim-Yalcin M, Demirkesen I et al (2024) Improving physicochemical and nutritional attributes of rice starch through green modification techniques. Food Chem 458:140212

Roy P, Orikasa T, Okadome H et al (2011) Processing conditions, rice properties, health and environment. Int J Environ Res Public Health 8:1957–1976

Sangpring Y, Fukuoka M, Ratanasumawong S (2015) The effect of sodium chloride on microstructure, water migration, and texture of rice noodle. LWT – Food Sci Technol 64:1107–1113

Sankar GS, Sankar SS, Subrata S et al (2018) Use of a potential probiotic, lactobacillus plantarum 17, for the preparation of a rice-based fermented beverage. Front Microbiol 9:1–11

Sawatari Y, Sugiyama H, Suzuki Y et al (2005) Development of fermented instant Chinese noodle using *Lactobacillus plantarum*. Food Microbiol 22:539–546

Sengupta A, Chakraborty I, Indira G et al (2023) An insight into the physicochemical characterisation of starch-lipid complex and its importance in food industry. Food Rev Int 39:4198–4212

Shi P, Zhao Y, Qin F et al (2023) Understanding the multi-scale structure and physicochemical properties of millet starch with varied amylose content. Food Chem 410:135422

Shoukat R, Cappai M, Pilia L et al (2025) Rice starch chemistry, functional properties, and industrial applications: a review. Polymers 17:110

Soman V, Suresh A, Remani AK et al (2024) A holistic review on diverse lipid biomarkers as environmental indicators: extraction and analysis from sediments to microbial communities. Geomicrobiol J 41:891–909

Song Y, Lin J, Luo L et al (2024) Solid-state fermentation of *Limosilactobacillus fermentum* and lysine addition to improve the cooking and eating quality of high-fiber rice noodles. J Cereal Sci 118:103959

Steensels J, Verstrepen KJ (2014) Taming wild yeast: potential of conventional and nonconventional yeasts in industrial fermentations. Ann Rev Microbiol 68:61–80

Sun X, Sun Z, Saleh ASM et al (2021) Understanding the granule, growth ring, blocklets, crystalline and molecular structure of normal and waxy wheat A- and B- starch granules. Food Hydrocoll 121:107034

Sun X, Yu L, Xiao M et al (2025) Exploring core fermentation microorganisms, flavor compounds, and metabolic pathways in fermented Rice and wheat foods. Food Chem 463:141019

Tantala J, Meethongchai S, Suethong W et al (2022) Mold-free shelf-life extension of fresh rice noodles by synergistic effects of chitosan and common food preservatives. Food Control 133:108597

Taşeri L, Aktaş M, Şevik S et al (2018) Determination of drying kinetics and quality parameters of grape pomace dried with a heat pump dryer. Food Chem 260:152–159

Thammathongchat S, Fukuoka M, Watanabe H (2024) An innovative noodle: gelatinized at the core, leaving the surface ungelatinized. J Food Eng 70:27–33

Tian J, Qin L, Zeng X et al (2023a) The role of amylose in gel forming of rice flour. Foods 12:1210

Tian Y, Wang Y, Zhong Y et al (2023b) Interfacial catalysis during amylolytic degradation of starch granules: current understanding and kinetic approaches. Molecules 28:3799

Tian Y, Petersen BL, Liu X et al (2024) Characterization of different high amylose starch granules. Part II: structure evolution during digestion and distinct digestion mechanisms. Food Hydrocoll 149:109593

Tong C, Chen Y, Tang F et al (2014) Genetic diversity of amylose content and RVA pasting parameters in 20 rice accessions grown in Hainan, China. Food Chem 161:239–245

Tracy BP, Jones SW, Fast AG et al (2012) Clostridia: the importance of their exceptional substrate and metabolite diversity for biofuel and biorefinery applications. Curr Opin Biotechnol 23:364–381

Ukkunda NS, Santhoshkumar P, Paranthaman R et al (2024) X-ray diffraction and its emerging applications in the food industry. Crit Rev Food Sci Nutr. https://doi.org/10.1080/10408398.2024.2395487

Valerio F, Conte A, Biase MD et al (2017) Formulation of yeast-leavened bread with reduced salt content by using a *Lactobacillus plantarum* fermentation product. Food Chem 221:582–589

Wan J, Huang W, Zhong J et al (2011) Effects of lab fermentation on physical properties of oat flour and its suitability for noodle making. Cereal Chem 88:153–158

Wang A, Xiao T, Xi H et al (2022) Edible qualities, microbial compositions and volatile compounds in fresh fermented rice noodles fermented with different starter cultures. Food Res Int 156:111184

Wang L, Guo M, Zhang Y et al (2024a) Effects of soaking treatment on water distribution of rice grains, starch characteristics and eating quality of wet rice noodles. Int J Biol Macromol 278:134621

Wang Y, Ou X, Al-Maqtari QA et al (2024b) Evaluation of amylose content: structural and functional properties, analytical techniques, and future prospects. Food Chem: X 24:101830
Wei P, Fang F, Liu G et al (2022) Effects of composition, thermal, and theological properties of rice raw material on rice noodle quality. Front Nutr 9:1003657
Weiling M, Yingjiao P, Chunmin G et al (2024) Effects of different fermentation processes on cooking and eating quality of brown rice noodles. LWT 208:116706
Xiong G, Lin L, Liu C (2023) Effects of solid-state fermentation with Lactobacillus fermentum and lysine addition cooking, eating, and storage quality of brown rice noodles. Paper presented at the twentieth annual conference of the Chinese society for food science and technology, changsha, hunan, China
Xu Y, Jin Y, Su J et al (2021) Changes in the nutritional value, flavor, and antioxidant activity of brown glutinous rice during fermentation. Food Biosci 43:101273
Xue C, Sakai N, Fukuoka M (2008) Use of microwave heating to control the degree of starch gelatinization in noodles. J Food Eng 87:357–362
Yang Y, Tao WY (2008) Effects of lactic acid fermentation on FT-IR and pasting properties of rice flour. Food Res Int 41:937–940
Yang S, Dhital S, Zhang M-N et al (2022) Structural, gelatinization, and rheological properties of heat-moisture treated potato starch with added salt and its application in potato starch noodles. Food Hydrocoll 131:107802
Yi C, Yang Y, Zhou S et al (2017) Role of lactic acid bacteria in the eating qualities of fermented rice noodles. Cereal Chem 94:349–356
Yi C, Zhu H, Tong L et al (2019) Volatile profiles of fresh rice noodles fermented with pure and mixed cultures. Food Res Int 119:152–160
Yi C, Zhu H, Bao J et al (2020a) The texture of fresh rice noodles as affected by the physicochemical properties and starch fine structure of aged paddy. LWT 130:109610
Yi C, Zhu H, Yang R et al (2020b) Links between microbial compositions and volatile profiles of rice noodle fermentation liquid evaluated by 16S rRNA sequencing and GC-MS. LWT 118:108774
Yi C, Zhu H, Zhang Y et al (2021) The role of indica starch in the mechanism of formation of fresh rice noodles. J Cereal Sci 99:103212
Yu Y, Ge L, Zhu S et al (2015) Effect of presoaking high hydrostatic pressure on the cooking properties of brown rice. J Food Sci Technol 52:7904–7913
Yu T, Jing S, Jiaxin et al (2024) Effects of milling methods on rice flour properties and rice product quality: a review. Rice Sci 31:33–46
Yvon M, Rijnen L (2001) Cheese flavour formation by amino acid catabolism. Int Dairy J 11:185–201
Zeng ZC, Wang YH, Xu GR et al (2023) Peroxidase inactivation by cold plasma and its effects on the storage, physicochemical and bioactive properties of brown rice. Food Biosci 52:102383
Zhang DQ, Ye YJ, Tan B (2022) Comparative study of solid-state fermentation with different microbial strains on the bioactive compounds and microstructure of brown rice. Food Chem 397:133735
Zhang X, Wang Z, Wang L et al (2023) Structural support of zein network to rice flour gluten-free dough: rheological, textural and thermal properties. Food Hydrocoll 141:108721
Zhang J, Zhao F, Li C et al (2024) Acceleration mechanism of the rehydration process of dried rice noodles by the porous structure. Food Chem 431:137050
Zhong Y, Li Z, Qu J et al (2021) Relationship between molecular structure and lamellar and crystalline structure of rice starch. Carbohydr Polym 258:117616
Zhu Y, Luo Y, Wang P et al (2016) Simultaneous determination of free amino acids in Pu-erh tea and their changes during fermentation. Food Chem 194:643–649
Zhu X, Guo X, Zhu X (2023) Effect of sorbitol on the *in vitro* starch digestibility in semi-dried black highland barley noodles. Int J Biol Macromol 23:123959
Zhu Z, Bao S, Du K et al (2024) Controlling high amylose corn starch modulates the structural features and digestive properties of rice noodles. LWT 209:116731

Chapter 4
The Raw Materials in Rice Noodles and Their Roles

Contents

Rice-based noodles are typically made from basic ingredients such as water and rice flour (Li et al. 2021), with additional components like oil, salt, hydrocolloids, polyphosphates, preservatives, and functional ingredients to enhance the final product's quality. The selection of ingredients with suitable processing and functional properties is essential for producing high-quality noodles (Kraithong et al. 2023). Several approaches have been explored to improve the quality of rice noodles, simplifying preparation and ensuring the formation of a cohesive dough structure (Huang et al. 2024). Previous research has demonstrated that the inclusion of food additives, such as enzymes, proteins, emulsifiers, hydrocolloids, and modified starches, can significantly improve the texture and overall quality of rice noodles (Kasunmala et al. 2020). Optimization and modification of these ingredients further contribute to enhancing noodle quality (Kasunmala et al. 2020).

Texture is a key attribute of cooked noodles, as it significantly impacts the consumer's mouthfeel and overall quality of the rice noodles, including characteristics such as hardness and springiness (Wang et al. 2022). The addition of additives and other process variables may influence the cohesiveness of the noodle dough and the structure of noodle strands, and thus thereby affecting the textural properties of the final product. A study by Kasunmala et al. (2020) revealed that rice noodles extruded

B. L. Tan, M. E. Norhaizan, *Rice Noodles: Quality Improvement, Processing Techniques and Nutritional Aspects*, SpringerBriefs in Food, Health, and Nutrition,
https://doi.org/10.1007/978-3-031-98649-9_4

at 75 °C were the best extrusion temperature in terms of texture profile analysis and cooking quality. The data demonstrated that dough temperature enhanced the cooking quality of rice noodles but did not show a significant effect on the textural quality of rice noodles. Compared to the noodles prepared from red rice flour, noodles prepared from white rice flour demonstrated a significantly high textural characteristic and cooking quality (Kasunmala et al. 2020).

4.1 Rice Flour

Rice flour is a versatile ingredient with a significant role in a wide range of culinary traditions, particularly in the production of rice noodles (Shen et al. 2021). As a staple in several Asian cuisines, rice noodles are cherished for their flavor, texture, and versatility in various dishes (Bangar et al. 2023). Rice flour is produced by finely milling rice kernels, which can be derived from either white rice or brown rice, each contributing distinct characteristics to the final product (Yu et al. 2024). During the milling process, rice grains have their husk removed, leaving behind the starchy interior, which is then ground into a fine powder. Rice flour is highly digestible, making it a fundamental food consumed regularly in various cuisines (Bao and Bergman 2024). Rice flour consists of 78% carbohydrates, primarily starch, which is composed of amylose and amylopectin. Amylose is a linear polymer, while amylopectin is branched, and the ratio of these two starches plays a significant role in determining the texture and firmness of rice flour-based products. A lower amylose content results in a softer and more tender texture, whereas higher amylose levels contribute to chewier and firmer noodles (Huang et al. 2023). In addition to carbohydrates, rice flour contains approximately 7% protein and 0.4%–0.8% fat, offering substantial amounts of riboflavin, niacin, and thiamine (Ahmed et al. 2016). Rice flour is versatile and can be incorporated into a variety of food products, such as gluten-free diets, salad dressings, puddings, meat products, beverages, cakes, noodles, and bread (Bresciani et al. 2021). Organic rice flour has gained popularity as a key ingredient in health foods due to growing consumer concerns about health. It is considered a safer and healthier alternative to non-organic flour, which may contain chemical fertilizers, herbicides, pesticides, and irradiation (Kraithong et al. 2018).

Gluten is a key protein in wheat flour, providing the structural integrity essential for wheat-based noodles (Zhang et al. 2022). However, rice flour lacks gluten, and rice starch becomes the dominant component influencing the texture, cooking tolerance, and mouthfeel of rice noodles during chewing (Zhu et al. 2023). Rice noodles are prepared by combining rice flour with water to form a batter or dough, tailored to the desired noodle texture and thickness. The production process varies depending on the type of rice noodles, such as vermicelli or flat noodles (Low et al. 2020). For traditional flat rice noodles, a water and rice flour mixture is spread into thin sheets and steamed. These steamed sheets are then cut into specific widths, resulting in the flat noodles commonly used in various dishes. In contrast, rice vermicelli, which is thin and round noodles, are created by extruding rice flour dough through small holes to form long, slender strands (Mahmud et al. 2023).

Rice flour is fundamental in the noodle-making process, providing both texture and structure to the final product. During heating, the starches in rice flour undergo gelatinization, forming a gel-like consistency that helps the noodles maintain their shape while ensuring they remain tender and flexible (Rashwan et al. 2024). Variations in the properties of rice flour, such as amylose and amylopectin content, significantly influence the overall quality of the noodles. Unlike wheat-based noodles, rice noodles lack gluten, which results in reduced elasticity. The absence of gluten imparts rice noodles with their characteristic smooth and slippery texture, making them particularly well-suited for absorbing flavors in soups and broths (Bangar et al. 2023). These unique textural properties contribute to the widespread popularity of rice noodles in diverse culinary applications.

Rice noodles are a staple food in Asian cuisine, offering a neutral flavor that readily absorbs the tastes of broths and sauces, making them highly versatile for various culinary applications. From light and refreshing salads to hearty soups, rice noodles adapt seamlessly to a wide range of dishes. The gluten-free nature of rice flour makes rice noodles an excellent alternative for individuals with gluten sensitivity or celiac disease (Park and Kim 2023). In addition, brown rice flour, which retains the germ and bran of the rice grain, can be used to produce rice noodles with higher fiber content, thereby adding nutritional benefits. Brown rice noodles are characterized by a slightly coarser texture and a more robust flavor compared to those made from white rice flour, catering to diverse tastes and dietary preferences (Murali and Sukumar 2024).

4.2 Water

Water is the second predominant raw ingredient utilized for the production of rice noodles. Water plays a central role in facilitating biochemical and physical interactions in foods. However, its significance is often underestimated or overlooked. Typically, about 30%–35% of water is used in flour production to produce good dough (Hou 2001). Indeed, it is a crucial material necessary for gluten formation. It provides viscoelastic properties to dough during the processing of rice noodles. Water affects the noodle texture formation, sheeting process, gelatinization of starches, and hydration of flour. In addition, water is required for the physical and biochemical interaction that occurs in foods. During processing, the water content can be adjusted to control dough strength, significantly affecting machinability in molded products and ensuring the desired texture in the final product (Hatcher et al. 1999). In commercial noodle manufacturing, the water content is determined by the ingredients in the formula, the processing equipment, and the desired characteristics of the final rice noodle product (Li et al. 2021).

In the initial step of the production of rice noodles, water is added to the rice flour to form a batter or dough. The amount of water is crucial, as it results in the consistency of the mixture, which subsequently affects the texture of the final products. For instance, for producing rice vermicelli, the looser batter is preferred which is necessitating more water (Zhang et al. 2023). Water also serves as a medium that allows the rice flour to be uniformly hydrated. This hydration is essential for the starches in flour to swell and lead the gelatinization when heat is applied. Therefore, proper hydration is necessary to ensure the starch granules are evenly distributed within the batter or dough, which is essential for achieving a consistent texture in the noodles. Indeed, imbalanced water content can cause an undesirable mouthfeel, breakage, and uneven texture in the final product (Alemu 2023).

The most significant function of water in the production of rice noodles is the gelatinization of starch (Geng et al. 2023). Starch gelatinization is a process in which the starch granules in rice flour absorb water and swell when heating (Hu et al. 2020). This process transforms the mixture from a grainy or powdery texture into a cohesive and smooth mass that can be shaped into noodles. During gelatinization, the presence of heat and water leads the starch molecules in the rice flour to absorb water and break down. This leads to the formation of a gel-like structure, which is critical for the noodles to hold their shape. In the absence of sufficient water, the starches cannot gelatinize properly, causing the noodles crumbly, brittle, and unable to maintain their structure during cooking (Biliaderis 2009). The temperature of water, in which the gelatinization occurred, is also affected by the amount of water present. With adequate water, gelatinization can happen at a low temperature, making the process more efficient and ensuring the noodles achieve the desired texture (Obadi et al. 2021). When the water is inadequate, it can cause increase in gelatinization temperature, which potentially leads to poor texture and uneven cooking.

Water hardness, caused by impurities and dissolved salts, can affect the quality of the final product, primarily due to the presence of calcium, magnesium cations, sulfates, bicarbonates, and other minerals (Tan et al. 2020). In rice noodle production, water plays a significant role in determining mouthfeel and texture, which are essential to the culinary appeal. The right amount of water in the batter or dough ensures a smooth, cohesive texture that is neither too firm nor too soft. Proper hydration during cooking provides the desired elasticity, yielding noodles that are tender yet resilient with a slight chewiness (Zhang et al. 2021). Conversely, insufficient water can result in noodles that are dry, tough, and prone to breaking (Du et al. 2021), while excess water can make them mushy and lacking structural integrity. Therefore, the correct water content is crucial for achieving the ideal texture, allowing rice noodles to retain their shape and bite while absorbing flavors. Overall, water is an essential component throughout rice noodle production, influencing everything from dough formation to starch gelatinization and hydration of rice flour.

4.3 Salt

Salt is crucial in the production of rice noodles, particularly in texture. Rice flour is the predominant ingredient in rice noodles that lacks gluten, the protein that gives wheat-based noodles a chewy and elastic texture. Without gluten, rice noodles can break apart and brittle easily. Nevertheless, an addition of salt during dough-making process facilitates strengthening the noodles' structure (Obadi et al. 2022). Salt binds to the proteins in rice flour, thus promoting a better gelation process, which is critical for the final texture of noodles (Weng et al. 2020). The ions in salt create more cohesive and firmer dough, and thus making the noodles less prone to breaking and have a more desirable chewiness (Fan et al. 2020). This role is crucial in rice noodles that are delicate and thin, in which texture is a key quality indicator.

In addition to texture enhancement, salt also plays a crucial role in controlling the water activity in rice noodles (Zhao et al. 2022). Water activity is the amount of free water available in the rice noodles, which influences the texture and shelf-life (Obadi et al. 2023). By decreasing the water activity, salt facilitates in suppression of microorganism proliferation, thus extending the shelf-life of partially dried and dried rice noodles (Obadi et al. 2023).

Salt is a well-known flavor enhancer (Sun et al. 2022). In rice noodles, salt increases the natural and subtle flavors of the rice while enhancing the overall taste of the dish. Notably, salt can enhance the inherent sweetness of rice and balance the flavor when the noodles are mixed with other ingredients including sauces, meats, and vegetables (Dong et al. 2024).

The intake of rice noodles, a gluten-free product, has increased in recent decades due to the spreading of celiac disease worldwide (Rao et al. 2024). Good rice noodle has a soft texture and white. Nonetheless, several dried rice noodle shows a hard texture after cooking (rehydration). Therefore, the manufacturers add salt to dried rice noodles to increase the elasticity and prevent the hardness of the cooked rice noodles. Salt is a primary material in the production of noodles, contributing about 1%–3% of the total weight of the noodles. It plays a crucial role in enhancing the properties of dough sheets, especially the noodles with a high water absorption index. It is well-recognized that the presence of salt in wheat noodles influences the noodle microstructure by increasing the development of the gluten network (Wang et al. 2024). The previous study has reported that salt enhances the development of a gluten network in wheat noodles thereby leading to a uniform gluten network and smooth structure (Dexter et al. 1979). Subsequently, this structure produces a high resistance and strong wheat noodles. Consistent with the study reported by Dexter et al. (1979), Salovaara (1982), and Lynch et al. (2009) found that salt improves the gluten network of bread dough. In rice noodles, scanning electron micrographs of the cross-section of dried rice noodles revealed that the inner structure was compact and dense (Xie et al. 2023). Subsequently, this structure gives a hard texture to the rice noodles. A study by Sangpring et al. (2015) evaluated NaCl in relation to the texture of cooked rice noodles, water migration in rice noodles during cooking, and microstructure of dried rice noodles. The study showed that NaCl decreased the packing of starch lumps in the rice noodles and improved the development of the protein network. Intriguingly, Sangpring et al. (2015) found that water migration in rice noodles containing NaCl occurred more slowly compared to the control. This was attributed to the reduced depth, size, and number of pores in the dried rice noodles with NaCl, which were diminished due to the formation of a protein network. The study further demonstrated that NaCl decreased the tensile strength and increased the extensibility of the cooked rice noodles. Increased concentration of NaCl enhanced the cooking loss (Sangpring et al. 2015). Besides dough sheets, salt also improves the taste and texture. It acts as a flavor enhancer in a variety of foods. Salt can also prevent the oxidative reactions and damage caused by high temperatures and humidity, and thus may extend the shelf-life of the noodles. However, a high concentration of salts in noodles may lead to slow drying (Zhang et al. 2024).

4.4 Oil

Oils have several functions in rice noodles, including flavor enhancement, texture improvement to moisture retention (Huang et al. 2022). They contributed not only to the visual appeal and nutritional value but also culinary qualities of the dish. Hence, understanding the role of oils in rice noodles can increase the cooking experience, satisfying meals and allowing for the creation of delicious meals. Selecting the right type of oil and using it appropriately can contribute to a significant outcome of rice noodle dishes, and ensure they are both flavorful and enjoyable.

One of the primary functions of oils in rice noodles is to enhance flavor. Oils, such as sesame oil, peanut oil, or vegetable oil, are often used in the cooking process or as a finishing touch. These oils impart distinct flavors that complement the subtle taste of rice noodles. For instance, sesame oil adds a nutty aroma, while peanut oil contributes a rich, slightly sweet flavor (Xu et al. 2024). The choice of oil can significantly influence the overall taste profile of the dish, making it more appealing to the palate.

Oils play a vital role in improving the texture of rice noodles. Rice noodles can become sticky and clump together when cooked (Low et al. 2020). Oil facilitates in coating the noodles thus preventing rice noodles from sticking and ensuring a slippery and smooth texture during the cooking process. Indeed, this is crucial in dishes such as stir-fries or salads, to ensure the noodles maintain the integrity and not become a mushy mass. In addition, the oil provides a barrier to separate the noodles and improves the overall mouthfeel of the noodles (Obadi et al. 2021).

In addition to enhancing flavor and texture, oils also facilitate moisture retention in rice noodles. Incorporating oil facilitates sealing in moisture, keeping the noodles juicy and tender. This is crucial in dishes that are served at room temperature or warm, as it prevents the noodles from becoming unpalatable or from drying out. The oil serves as a protective layer and ensures the noodles remain moist (Huang et al. 2024).

4.5 Hydrocolloids and Polyphosphates

Hydrocolloids are a group of polysaccharides commonly utilized in food products to enhance their ability to alter viscosity, texture, and water retention (Ali et al. 2024). Hydrocolloids, for example, agar, carboxymethyl cellulose (CMC), xanthan gum, and guar gum are commonly used in rice noodles production. Hydrocolloids play an important role in rice noodles such as stabilizing noodle structure, enhancing water retention, and improving texture (Kraithong et al. 2023). Hydrocolloids can improve the overall quality of noodles. Guar gum is often utilized in the production of rice noodles. The amount addition of hydrocolloids ranges between 0.2% and 0.5% of the total weight of the flour.

Hydrocolloids facilitate improving the firmness and elasticity of rice noodles by forming a gel-like network that mimics the structural properties of gluten in wheat-based noodles (Qazi et al. 2011). The network can help in the binding of starch granules and water, thus reducing stickiness and better chewiness. Hydrocolloids also enhance the water-holding capacity of rice noodles, which is crucial in maintaining the texture during cooking (Tan et al. 2018b). By retaining moisture, hydrocolloids can prevent the noodles from becoming brittle or too dry. They also decrease cooking loss by minimizing the leaching of starch into the cooking water, thus preserving the texture and nutritional quality of noodles.

Starch retrogradation is a common issue in rice noodles, and thus leading to textural changes including syneresis (water expulsion) or hardening during storage (Bangar et al. 2023). Hydrocolloids can suppress starch retrogradation by binding to starch molecules, and hence maintaining the quality and extending the shelf life of noodles.

Polyphosphates are a group of food additives, consisting of linked phosphate units. They are utilized in a wide spectrum of food applications, such as seafood preservation (Talab et al. 2022) and meat processing, as well as rice noodle production. The primary role of polyphosphates in rice noodles such as extending shelf life (Junior et al. 2024), improve water-binding capacity, and enhance texture.

Polyphosphates facilitate in enhancing the texture of rice noodles by alteration of starch and protein interactions within the noodle matrix (Mahmud et al. 2023). They work by chelating calcium ions, which enhances the elasticity and firmness of the noodles and strengthens the noodle structure.

Polyphosphates prevent dryness, facilitate in maintaining moistness, and improve the water-binding capacity of rice noodles (Bangar et al. 2023). Polyphosphates act as pH regulators, which can affect the color and texture of noodles. By adjusting the pH, polyphosphates can also prevent undercooking or overcooking and facilitate controlling the cooking time.

Polyphosphates exert antioxidant properties that can facilitate in preventing oxidative rancidity (Gumus-Bonacina et al. 2024), which is especially crucial in noodles containing fat. Moreover, polyphosphates can suppress microbial growth, and thus ensuring the quality and safety during storage and extending the shelf life of rice noodles (Mahmud et al. 2023).

The addition of mucilaginous materials and gums significantly improved the textural and cooking qualities of rice noodles. Xanthan gum-incorporated noodles imparted the best textural and cooking quality characteristics compared to the mucilage materials incorporated in noodles and the other gums. A study reported by Kasunmala et al. (2020) showed that incorporating mucilage extract of Neolitsea *cassia* leaves into noodles showed higher textural and cooking quality compared to the Gum Arabic incorporated noodles. The study further revealed that increasing the concentration of gum did not significantly influence the textural qualities and cooking tolerance of cooked rice noodles, suggesting the textural of rice noodles can be enhanced by affecting the extrusion temperature concomitantly with the binding materials (Kasunmala et al. 2020).

4.6 Preservatives

The preservatives in rice noodles play a crucial role in maintaining quality and extending shelf life (Li et al. 2021). Preservatives suppress the proliferation of yeasts, molds, and bacteria that make them unsafe to eat or could spoil the noodles. Preservatives maintain the desired texture of rice noodles by preventing the rice noodles from becoming sticky or too hard. It also prevents oxidation which could lead to off-flavors and spoilage in rice noodles. Antioxidant, for instance, ascorbic acid is often used as a preservative. Preservatives allow rice noodles to be stored for a longer period without losing quality. Some preservatives prevent the action of enzymes that can result in undesirable changes in the texture, flavor, or color of rice noodles during storage (Yi et al. 2020).

4.7 Functional Ingredients

Functional ingredients are components added to food products that provide health benefits (Galanakis 2021). Functional ingredients can improve texture and enhance nutritional value promoting overall health (Gupta et al. 2023). In general, rice noodles are made with water and rice flour, which are relatively low in vitamins and fibers (Li et al. 2021). By incorporating functional ingredients, rice noodles can be transformed into health-promoting and nutritious food products (Abid et al. 2024). This has become increasingly crucial as consumers seek healthier food that fits into a wide range of dietary requirements.

Dietary fibers are commonly added to rice noodles to enhance the fiber content, which is relatively low in rice flour (Liu et al. 2021). Fibers such as inulin or psyllium husk can improve digestive health by enhancing bowel regularity and increasing stool bulk (Tan et al. 2018a). Fiber-enriched rice noodles can have a lower glycemic index, which is beneficial in modulating blood glucose levels (Luo et al.

2024). In addition, fiber can enhance the mouthfeel and texture of rice noodles, leading to more elasticity and firmness.

Rice noodles are relatively low in protein. Incorporating functional protein sources in rice noodles can improve the nutritional value. Pea protein or soy protein isolates are excellent plant-based proteins that can be incorporated into rice noodles (Liang et al. 2023). In addition to being high in protein, chia seeds and quinoa also provide fiber and fatty acids. These ingredients can balance the macronutrients in rice noodles, transforming the rice noodles from carbohydrate-based food to a more balanced meal alternative food product. Fortification of rice noodles with minerals and vitamins is another option to improve the nutritional quality (Meenu et al. 2023). Iron and calcium are widely added to improve bone health, particularly in populations that consume rice-based products as a staple food (Rybicka and Gliszczyńska-Świgło 2017).

References

Abid J, Ahmed S, Xia T et al (2024) Rice as a nutritional grain: examining its role in healthy products and disease prevention. Food Rev Int 40:3055–3078

Ahmed I, Qazi IM, Li Z et al (2016) Rice noodles: materials, processing and quality evaluation. Proc Pak Acad Sci: B Life Environ Sci 53:215–238

Alemu T (2023) Texture profile and design of food product. J Agri Horti Res 6:272–281

Ali TM, Shaikh M, Haider S et al (2024) Chapter 2 – Significance of hydrocolloids in the formation of gluten-free pasta. In: Gull A, Nayik GA, Brennan C (eds) Development of gluten-free pasta. Academic, London, pp 19–44

Bangar SP, Ali NA, Olagunju AI et al (2023) Starch-based noodles: current technologies, properties, and challenges. J Texture Stud 54:21–53

Bao J, Bergman CJ (2024) Chapter 12 – Rice flour and starch functionality. In: Nilsson L (ed) Starch in food: structure, function, and applications, 3rd edn. Woodhead Publishing, Cambridge, pp 275–307

Biliaderis CG (2009) Chapter 8 – Structural transitions and related physical properties of starch. In: BeMiller J, Whistler R (eds) Starch: chemistry and technology, 3rd edn. Academic, London, pp 293–372

Bresciani A, Pagani MA, Marti A (2021) Rice: a versatile food at the heart of the Mediterranean diet. In: Boukid F (ed) Cereal-based foodstuffs: the backbone of Mediterranean cuisine. Springer, Cham, pp 193–229

Dexter JE, Matsuo RR, Dronzek BL (1979) A scanning electron microscopy study of Japanese noodles. Cereal Chem 56:202–209

Dong K, Wang Q, Li X et al (2024) An effective means to improve the flavor quality of traditional fermented sour meat: the salt reduction strategy. LWT 209:116726

Du J, Li Q, Obadi M et al (2021) Quality evaluation systems and methods of the whole making process of Asian noodles: a review. Food Rev Int 39:3830–3857

Fan H, Fu F, Chen Y et al (2020) Effect of NaCl on rheological properties of dough and noodle quality. J Cereal Sci 93:102936

Galanakis CM (2021) Functionality of food components and emerging technologies. Foods 10:128

Geng D-H, Tang N, Zhang X et al (2023) Insights into the textural properties and starch digestibility on rice noodles as affected by the addition of maize starch and rice starch. LWT 173:114265

Gumus-Bonacina CE, Mcclements DJ, Decker EA (2024) Replacing animal fats with plant-based lipids: challenges and opportunities. Curr Opin Food Sci 58:101193

Gupta A, Sanwal N, Bareen MA et al (2023) Trends in functional beverages: functional ingredients, processing technologies, stability, health benefits, and consumer perspective. Food Res Int 170:113046

Hatcher DW, Kruger JE, Anderson MJ (1999) Influence of water absorption on the processing and quality of oriental noodles. Cereal Chem 76:566–572

Hou G (2001) Oriental noodles. Adv Food Nutr Res 43:143–193

Hu W-X, Chen J, Zhao J-W et al (2020) Effect of the addition of modified starch on gelatinization and gelation properties of rice flour. Int J Biol Macromol 153:26–35

Huang S, Chi C, Li X et al (2022) Understanding the structure, digestibility, texture and flavor attributes of rice noodles complexation with xanthan and dodecyl gallate. Food Hydrocoll 127:107538

Huang H, Li Y, Zeng J et al (2023) Comparative quality evaluation of physicochemical and amylose content profiling in rice noodles from diverse rice hybrids in China. Agriculture 13:140

Huang G, McClements DJ, He K et al (2024) Review of formation mechanisms and quality regulation of chewiness in staple foods: rice, noodles, potatoes and bread. Food Res Int 187:114459

Junior WJFL, Santinello D, Mohammadzadeh S et al (2024) Polyphosphate in food systems: their roles and applications in foods and contribution to sustainable processing practices. Trends Food Sci Technol 152:104696

Kasunmala IGG, Navaratne SB, Wickramasinghe I (2020) Effect of process modifications and binding materials on textural properties of rice noodles. Int J Gastron Food Sci 21:100217

Kraithong S, Lee S, Rawdkuen S (2018) Physicochemical and functional properties of Thai organic rice flour. J Cereal Sci 79:259–266

Kraithong S, Theppawong A, Lee S et al (2023) Understanding of hydrocolloid functions for enhancing the physicochemical features of rice flour and noodles. Food Hydrocoll 142:108821

Li C, You Y, Chen D et al (2021) A systematic review of rice noodles: raw material, processing method and quality improvement. Trends Food Sci Technol 107:389–400

Liang J-N, Nargotra P, Li X-Y et al (2023) Evaluation of wheat noodles supplemented with soy protein isolate for nutritional, textural, cooking attributes and glycemic index. Appl Sci 13:7772

Liu T, Wang K, Xue W et al (2021) In vitro starch digestibility, edible quality and microstructure of instant rice noodles enriched with rice bran insoluble dietary fiber. LWT 142:111008

Low YK, Effarizah ME, Cheng LH (2020) Factors influencing rice noodles qualities. Food Rev Int 36:781–794

Luo X, Chang X, Zhuang K et al (2024) Feasible methods to control starch digestibility: strategies for reducing glycemic index of rice noodles. Trends Food Sci Technol 149:104536

Lynch EJ, Dal Bello F, Sheehan EM et al (2009) Fundamental studies on the reduction of salt on dough and bread characteristics. Food Res Int 42:885–891

Mahmud N, Islam J, Tahergorabi R (2023) Noodles. In: Shah MA, Valiyapeediyekkal SK, Mir SA (eds) Cereal-based food products. Springer, Cham, pp 221–252

Meenu M, Padhan B, Zhou J et al (2023) Fortification of rice noodles with minerals and vitamins is another option to improve the nutritional quality. Food Rev Int 39:6162–6198

Murali H, Sukumar M (2024) Assessing the influence of partial substitution of wheat flour by brown rice flour, kale and Pithecellobium Dulce extract on noodle properties: a study of physical, nutritional, antioxidant, and sensory characteristics. Int J Res Eng Sci Mang 7:120–125

Obadi M, Zhang J, Shi Y et al (2021) Factors affecting frozen cooked noodle quality: a review. Trends Food Sci Technol 109:662–673

Obadi M, Zhang J, Xu B (2022) The role of inorganic salts in dough properties and noodle quality—a review. Food Res Int 157:111278

Obadi M, Li Y, Xu B (2023) Recent advances in extending the shelf life of fresh wet noodles: influencing factors and preservation technologies. J Food Sci 88:3626–3648

Park J, Kim H-K (2023) Rice-based gluten-free foods and technologies: a review. Foods 12:4110

Qazi IM, Rakshit SK, Tran T (2011) Effect of physico-chemical properties of tropical starches and hydrocolloids on rice gels texture and noodles water retention ability. Starch 63:558–569

Rao SJN, Majaw J, Prabhasankar P (2024) Chapter 8 – In-vitro starch, protein digestibility and glycemic response of gluten-free pasta. In: Gull A, Nayik GA, Brennan C (eds) Development of gluten-free pasta. Academic, London, pp 179–206

Rashwan AK, Younis HA, Abdelshafy AM et al (2024) Plant starch extraction, modification, and green application: a review. Environ Chem Lett 22:2483–2530

Rybicka I, Gliszczyńska-Świgło A (2017) Minerals in grain gluten-free products. The content of calcium, potassium, magnesium, sodium, copper, iron, manganese, and zinc. J Food Compos Anal 59:61–67

Salovaara H (1982) Sensory limitations to replacement of sodium with potassium and magnesium in bread. Cereal Chem 59:425–430

Sangpring Y, Fukuoka M, Ratanasumawong S (2015) The effect of sodium chloride on microstructure, water migration, and texture of rice noodle. LWT – Food Sci Technol 64:1107–1113

Shen S, Chi C, Zhang Y et al (2021) New insights into how starch structure synergistically affects the starch digestibility, texture, and flavor quality of rice noodles. Int J Biol Macromol 184:731–738

Sun X, Zhong K, Zhang D et al (2022) The enhancement of the perception of saltiness by umami sensation elicited by flavor enhancers in salt solutions. Food Res Int 157:111287

Talab AS, Genina ME, Anees FR (2022) Effectiveness of some cryoprotectant agents on chemical composition, physicochemical, microbiological and sensory quality criteria of some marine fish species. J Food Dairy Sci 13:89–93

Tan BL, Norhaizan ME, Liew W-P-P et al (2018a) Antioxidant and oxidative stress: a mutual interplay in age-related diseases. Front Pharmacol 9:1162

Tan H-L, Tan T-C, Easa AM (2018b) The use of selected hydrocolloids to enhance cooking quality and hardness of zero-salt noodles. Int J Food Sci Technol 53:1603–1610

Tan H-L, Tan T-C, Musa AM (2020) Effects of sodium chloride or salt substitutes on rheological properties and water-holding capacity of flour and hardness of noodles. Food Struct 26:100154

Wang Y-H, Zhang Y-R, Yang Y-Y et al (2022) Effect of wheat gluten addition on the texture, surface tackiness, protein structure, and sensory properties of frozen cooked noodles. LWT 161:113348

Wang J, Li Y, Guo X et al (2024) A review of the impact of starch on the quality of wheat-based noodles and pasta: from the view of starch structural and functional properties and interaction with gluten. Foods 13:1507

Weng Z-J, Wang BJ, Weng Y-M (2020) Preparation of white salted noodles using rice flour as the principal ingredient and the effects of transglutaminase on noodle qualities. Food Biosci 33:100501

Xie L, Zhou W, Zhao L et al (2023) Impact of okara on quality and *in vitro* starch digestibility of noodles: the view based on physicochemical and structural properties. Int J Biol Macromol 237:124105

Xu H, Liu H, Yin W et al (2024) Roasted sesame oil: a review on production technology, flavor chemistry, and other related issues. J Am Oil Chem Soc 102(4). https://doi.org/10.1002/aocs.12917

Yi C, Zhu H, Bao J et al (2020) The texture of fresh rice noodles as affected by the physicochemical properties and starch fine structure of aged paddy. LWT 130:109610

Yu T, Jing S, Jiaxin L et al (2024) Effects of milling methods on rice flour properties and rice product quality: a review. Rice Sci 31:33–46

Zhang Y, Dong J, Deng C et al (2021) Effect of glutinous rice flour supplementation on the properties of wheat flour and salted noodles. J Food Process Preserv 45:e15856

Zhang M, Ma M, Yang T et al (2022) Dynamic distribution and transition of gluten proteins during noodle processing. Food Hydrocoll 123:107114

Zhang M, Jia R, Ma M et al (2023) Versatile wheat gluten: functional properties and application in the food-related industry. Crit Rev Food Sci Nutr 63:10444–10460

Zhang J, Zhao F, Li C et al (2024) Acceleration mechanism of the rehydration process of dried rice noodles by the porous structure. Food Chem 431:137050

Zhao T-T, Guo X-N, Zhu K-X (2022) Effect of phosphate salts on the shelf-life and quality characteristics of semi-dried noodles. Food Chem 384:132481

Zhu D, Zheng X, Yu J et al (2023) Effects of starch molecular structure and physicochemical properties on eating quality of Indica rice with similar apparent amylose and protein contents. Foods 12:3535

Chapter 5
Factors Influencing Rice Noodles Qualities

Contents

5.1 Physical and Chemical Properties of Rice Flour

Rice noodle is a very popular traditional food with unique texture, taste, and flavor in Southeast Asian and East countries (Yi et al. 2021; Kasunmala et al. 2020; Fu et al. 2020). Rice noodles are commonly processed from indica rice. It has been categorized into frozen, dried, and fresh products based on their shapes and thicknesses (Fu 2008; Tong et al. 2015). The quality of rice noodles is tightly associated with the chemical and physical properties of rice flour. In general, starch is the predominant component in rice. Similar to other crop starch (Li et al. 2021; Zhang et al. 2021, 2022), the viscoelastic property of rice noodles mainly depends on the starch properties and structure (Sandhu et al. 2010; Kim et al. 2014; Tong et al. 2015). A study by Zhou et al. (2018) has demonstrated a significant correlation between sensory score and the amylose content of rice noodles. Therefore, amylose content was chosen as the sensitive index for the prediction of the quality of rice noodles. The data revealed that amylose content suitable for processing the pressed fresh rice noodles was in the range of 22.2–26.9%. Another study found that rice with amylose content in the range of 20–25% is essential for rice noodle production (Xuan et al. 2020). Indeed, rice

B. L. Tan, M. E. Norhaizan, *Rice Noodles: Quality Improvement, Processing Techniques and Nutritional Aspects*, SpringerBriefs in Food, Health, and Nutrition,
https://doi.org/10.1007/978-3-031-98649-9_5

with an amylose content of more than 25% or less than 20% is not suitable for the production of rice noodles. In this regard, it can be observed that the range of amylose content for the production of rice noodles was different by different researchers. The rice starch content should be selected based on the actual demand for rice noodle products.

Rehydration time is closely related to cooking time, while breakage rate and cooking loss are critical indicators of rice noodle cooking quality (Wei et al. 2022). A study by Wei et al. (2022) explored the rice noodles cooking qualities. The study found that Zhengui (ZG), Shuanggui (SG), and Guichao (GC) rice noodles showed no significant differences in rehydration time (15–16 min), which was significantly longer than that of Suimi (SM) (12 min). SM exhibited a broken rate approximately twice as high as that of ZG, SG, and GC, with no significant differences observed among the latter three. The cooking loss of rice noodles followed the order: GC < ZG < SG < SM. The study further demonstrated that GC and ZG showed the best cooking quality, characterized by low breakage rates and minimal cooking loss, followed by SG. In contrast, SM had the lowest cooking quality, with the highest breakage rate and cooking loss. Several previous studies have shown that noodles with higher amylose content typically exhibit shorter rehydration times, greater gel strength, and lower breakage rates and cooking losses (Lii and Chang 1981; Qazi et al. 2011). The study by Wei et al. (2022) further revealed that SM noodles exhibited a darker color, shorter segments, and relatively poor shape. This may be attributed to greater structural damage to the rice starch granules during the shucking and stripping processes. In terms of sensory evaluation of rice noodles, rice noodles made from GC, ZG, and SG showed significant improvements in color, shape, and smoothness compared to noodles made from SM. The firmness scores were demonstrated to correlate with the amylose content of rice (Wei et al. 2022). Firmness scores increased significantly as the amylose content of rice increased from 21.00% to 23.91%. The texture of cooked noodles significantly influences their sensory properties and overall consumer acceptance. When the amylose content of GC rice was 23.91%, the hardness of rice starch noodles reached 1628.67 g, which is significantly higher than that of noodles made from ZG, SG, and SM, which had hardness values of 1084.00 g, 1097.00 g, and 941.67 g, respectively. This difference is likely associated with amylose content. GC rice flour, with an amylose content of 23.91%, had a higher amylose level than ZG, SG, and SM, resulting in stronger gel strength in GC starch compared to the other three varieties. The physicochemical properties of rice flour significantly affected the quality of rice noodles. Amylose content showed a positive correlation with noodle hardness and a strong positive correlation with peak and trough viscosities, chewiness, peak load, and deformation at peak in rice starch noodles. Nonetheless, Adhesive strength exhibited a strong negative correlation with hardness, chewiness, peak load, and deformation at peak in rice starch noodles.

5.2 Effect of Components on Quality of Rice Noodle

Rice protein (endosperm protein) is another important component of rice. Its content varies significantly among different rice varieties and is influenced by growing conditions. The protein content in rice is markedly varies in different varieties growing in different environments. The gelatinization of rice flour was affected by the formation of disulfide bonds and network structure formed by the protein binding with water (Martin and Fitzgerald 2002). The rice protein can be indirectly influenced by the adaptability of rice processing, primarily by changing the gelatinizing properties of starch, such as retrogradation, extrusion performance, and heat resistance. Intriguingly, the protein in rice could suppress the water absorption and expansion of starch particles, and leading to the increased gelatinization peak temperature of rice flour compared to the starch alone. Furthermore, the protein content can be used as an auxiliary index to select the raw material of rice flour for the production of noodles. The lipid content in rice starch is very low, but it is tightly linked to the gel properties of rice (Wei et al. 2022). Lipids showed a greater effect on rheological and gelatinization properties compared to protein (Ibáñez et al. 2007). In general, the presence of lipids promotes the formation of a gel system and decreases the gelatinization heat enthalpy of starch. The complex generated by starch and lipids maintains the stability of the gel structure, decreases the dissolution of starch, and prevents amylose crystallization, and subsequently extends the shelf-life and suppresses the production of aging in the products (Wei et al. 2022).

Rice protein lacks of gluten, and thus starch granules are one of the building blocks of the rice noodles strings, for instance, structural properties and quality (Jayaprakash et al. 2022). Therefore, the physiochemical properties of rice starch may affect the final quality of the product. In general, different rice noodles are produced from different sources of raw rice materials (primarily referring to the early polished *indica* rice), which predominantly depends on the experiences of rice producers using regional rice varieties as raw materials. For instance, dry rice noodles are commonly made from Guichao varieties; while some rice varieties are combined with broken rice for snail rice noodles production. A study by Wei et al. (2022) evaluated the correlations between the qualities of rice noodles and the physicochemical properties of rice flour. The study found that the amylose content of the rice had a positive correlation toward the hardness of rice noodles as well as a strong positive correlation to the peak viscosities, peak load and deformation at peak of rice starch noodles, and chewiness.

Gluten-free noodle is the only alternative for people suffering from celiac disease which is inherited by a specific genetic nature, due to the gliadin fraction of wheat gluten protein (Singla et al. 2024). Hence, the development of gluten-free noodles is crucial but it is hard as it does not form a cohesive and uniform matrix to withstand cooking stress and thus contributing a clear bite of chewable noodles.

5.3 Effects of Fermentation on Rice Noodle Quality

Fermented rice noodles develop a unique flavor due to the activity of microorganisms and enzymes (Keatkrai and Jirapakkul 2010; Lu et al. 2005). A study by Geng et al. (2020) found that the flavor score of pure and pure inoculum-fermented rice noodles was higher than that of natural and natural inoculum-fermented rice noodles. The difference in flavor score is associated with the bacterial composition of the fermented rice slurry and the volatile compounds generated during fermentation. *Lactobacillus plantarum*, the dominant bacterium in pure and pure inoculum-fermented rice slurries, produced rice noodles with a pronounced fermented rice aroma. This finding aligned with the previous studies, which demonstrated that rice noodles fermented with pure starter cultures exhibit superior aroma characteristics compared to those fermented with mixed cultures (Yi et al. 2019). In addition, the relative content of aldehydes in pure and pure inoculum-fermented rice noodles was higher than in natural and natural inoculum-fermented ones (Geng et al. 2020). Aldehydes with five to ten carbon atoms have a low odor threshold and impart a desirable aroma to fermented rice noodles (Iizuka-Furukawa et al. 2017; Zhang et al. 2018). Hydrocarbons were also detected in pure and pure inoculum-fermented rice noodles; however, their high odor threshold means they contribute minimally to flavor (Zhang et al. 2017). Esters were prominent volatile compounds in natural and natural inoculum-fermented rice noodles. Previous studies suggest that low molecular weight esters possess fruity aromas, potentially enhancing the flavor of natural and natural inoculum-fermented rice noodles (Medeiros et al. 2006).

References

Fu BX (2008) Asian noodles: history, classification, raw materials, and processing. Food Res Int 41:888–902

Fu M, Sun X, Wu D et al (2020) Effect of partial substitution of buckwheat on cooking characteristics, nutritional composition, and *in vitro* starch digestibility of extruded gluten-free rice noodles. LWT 126:109332

Geng D-H, Liu L, Zhou S et al (2020) Effects of *Lactobacillus plantarum* inoculum on the fermentation rate and rice noodle quality. J Oleo Sci 69:1031–1041

Ibáñez AM, Wood DF, Yokoyama WH et al (2007) Viscoelastic properties of waxy and nonwaxy rice flours, their fat and protein-free starch, and the microstructure of their cooked kernels. J Agric Food Chem 55:6761–6771

Iizuka-Furukawa S, Isogai A, Kusaka K et al (2017) Identification of 4-mercapto-4-methylpentan-2-one as the characteristic aroma of sake made from low-glutelin rice. J Biosci Bioeng 123:209–215

Jayaprakash G, Bains A, Chawla P et al (2022) A narrative review on rice proteins: current scenario and food industrial application. Polymers 14:3003

Kasunmala IGG, Navaratne SB, Wickramasinghe I (2020) Effect of process modifications and binding materials on textural properties of rice noodles. Int J Gastron Food Sci 21:100217

Keatkrai J, Jirapakkul W (2010) Volatile profile of *khanom jeen*, Thai fermented rice noodles, and the changes during the fermentation process. ScienceAsia 36:46–51

Kim Y, Kee JI, Lee S et al (2014) Quality improvement of rice noodle restructured with rice protein isolate and transglutaminase. Food Chem 145:409–416
Li B, Zhang Y, Xu F et al (2021) Supra molecular structure of Artocarpus heterophyllus Lam seed starch prepared by improved extrusion cooking technology and its relationship with in vitro digestibility. Food Chem 336:127716
Lii CY, Chang SM (1981) Characterization of red bean (Phaseolus radiatus var. Aurea) starch and its noodle quality. J Cereal Sci 46:78–81
Lu ZH, Li LT, Min WH et al (2005) The effects of natural fermentation on the physical properties of rice flour and the rheological characteristics of rice noodles. Int J Food Sci Technol 40:985–992
Martin M, Fitzgerald MA (2002) Proteins in rice grains influence cooking properties. J Cereal Sci 36:285–294
Medeiros ABP, Pandey A, Vandenberghe LPS et al (2006) Production and recovery of aroma compounds produced by solid-state fermentation using different adsorbents. Food Technol Biotechnol 44:47–51
Qazi IM, Rakshit SK, Tran T (2011) Effect of physico-chemical properties of tropical starches and hydrocolloids on rice gels texture and noodles water retention ability. Starch-Starke 63:558–569
Sandhu KS, Kaur M, Mukesh (2010) Studies on noodle quality of potato and rice starches and their blends in relation to their physicochemical, pasting and gel textural properties. LWT-Food Sci Technol 43:1289–1293
Singla D, Malik T, Singh A et al (2024) Advances in understanding wheat-related disorders: a comprehensive review on gluten-free products with emphasis on wheat allergy, celiac and non-celiac gluten sensitivity. Food Chem Adv 4:100627
Tong LT, Gao XX, Lin LZ et al (2015) Effects of semidry flour milling on the quality attributes of rice flour and rice noodles in China. J Cereal Sci 62:45–49
Wei P, Fang F, Liu G et al (2022) Effects of composition, thermal, and theological properties of rice raw material on rice noodle quality. Front Nutr 9:1003657
Xuan Y, Yi Y, Liang H et al (2020) Amylose content and RVA profile characteristics of noodle rice under different conditions. Agron J 112:117–129
Yi C, Zhu H, Tong L et al (2019) Volatile profiles of fresh rice noodles fermented with pure and mixed cultures. Food Res Int 119:152–160
Yi C, Zhu H, Zhang Y et al (2021) The role of indica starch in the mechanism of formation of fresh rice noodles. J Cereal Sci 99:103212
Zhang Y, Song H, Li P et al (2017) Determination of potential off-flavour in yeast extract. LWT Food Sci Technol 82:184–191
Zhang H, Pu D, Sun B et al (2018) Characterization and comparison of key aroma compounds in raw and dry porcini mushroom *Boletus edulis* by aroma extract dilution analysis, quantitation and aroma recombination experiments. Food Chem 258:260–268
Zhang Y, Zuo H, Xu F et al (2021) The digestion mechanism of jackfruit seed starch using improved extrusion cooking technology. Food Hydrocoll 110:106154
Zhang Y, Xu F, Wang Q et al (2022) Effects of moisture content on digestible fragments and molecular structures of high amylose jackfruit starch prepared by improved extrusion cooking technology. Food Hydrocoll 133:108023
Zhou X, Peng C, Zhang Y (2018) Quality analysis of early indica rice cultivars and their suitability for processing of pressed fresh noodle. Food Sci 39:36–43

Chapter 6
Qualities and Nutritional Values Improvement of Rice Noodles

Contents

Rice noodles are the most consumed form of rice-based food product in Asia (Park and Kim 2023). Rice noodles known as *bee hoon*, are regarded as a popular and traditional food in China (Li et al. 2015). Rice noodles, known for their elastic and smooth texture, are a staple food widely consumed in Asia. With the growing consumer demand for healthy and high-quality foods, many food manufacturers and research institutions are focusing on whole grain products and are committed to developing new functional foods (Cáceres et al. 2019).

Starch is the largest component in fresh rice noodles. The texture of starch gels has a positive effect on the formation of fresh rice noodles structure (Yi et al. 2021). In recent years, consumers have placed greater emphasis on health benefits alongside taste preferences (Guiné et al. 2020). Research evidence indicates that amylopectin and amylose affect the formation of fresh rice noodles (Yi et al. 2021). The fresh rice noodles made from NR rice cultivar demonstrated an excellent fresh rice noodle texture. Intriguingly, the extracted starch from BX02, Zao 87, and Zhongzao 35 rice cultivars, with higher fractions of amylopectin (FrII) and lower fractions of amylose (FrI) or higher swelling power could not produce a consistent and dense fresh rice noodle structure (Yi et al. 2021).

The predominant ingredient of rice noodles is polished white rice flour which lacks certain nutritional values. The study by Muniandy and Gannasin (2019) demonstrated that rice noodles prepared with brown rice flour showed better nutrient

B. L. Tan, M. E. Norhaizan, *Rice Noodles: Quality Improvement, Processing Techniques and Nutritional Aspects*, SpringerBriefs in Food, Health, and Nutrition,
https://doi.org/10.1007/978-3-031-98649-9_6

content compared to 100% white rice flour. The data further revealed that 100% brown rice flour noodles showed lower carbohydrate content compared to 100% white rice flour noodles. This observation could be attributed to the presence of germ and bran in brown rice flour. The germ and bran of the whole grain are rich in fiber, lipids, and protein, and thus resulting in lower starch content in brown rice flour. However, white rice flour made from white rice grain, in which the rice bran has been removed, has higher starch content and resulting in increased carbohydrate levels in noodles made from 100% white rice flour (Muniandy and Gannasin 2019). Huang et al. (2021) reported that the yield of brown rice noodles was 12%–19% higher than that of white rice noodles. This finding suggests that the higher milling recovery rate of brown rice, compared to white rice, is retained during the production of rice noodles. The cooking loss rate and cooked break rate of brown rice noodles are higher compared to white rice noodles, implying that brown rice noodles had poorer cooking quality than white rice noodles. Nonetheless, both the cooking loss rate (about 12%) and cooked break rate (about 11%) of brown rice noodles remain within the acceptable range of less than 15% (Jiao et al. 2020). The lower cooking quality of brown rice noodles compared to white rice noodles could be attributed to the decreased paste viscosities in brown rice flour than white rice flour. A study by Bhattachrya et al. (1999) further supported that the cooking loss rate of rice noodles was inversely associated with the paste viscosities of rice flours. The paste properties of rice flour are tightly associated with the composition. The addition of rice bran into rice flours significantly decreased the paste viscosities (Saleh et al. 2014). Geng et al. (2020) also found that the final, trough, and peak viscosities were positively linked to the total starch content in rice flour. In this regard, the reduction of paste viscosities in brown rice flour than white rice flour was attributable to the decreased total starch content and retained rice bran. Furthermore, cooked brown rice noodles showed 21%–27% lower chewiness and hardness than cooked white rice noodles, despite the differences in resilience, cohesiveness, and springiness were inconsistent or not significant between cooked white and brown rice noodles. The inconsistent difference in resilience between cooked white and brown rice noodles across cultivars indicates that the texture quality of brown rice noodles could be enhanced via cultivar selection (Huang et al. 2021).

A study conducted by Qiao et al. (2022) evaluated the moisture migration, texture properties, and quality characteristics of fresh brown rice noodles in various storage processes at −18 °C (0–120 h), 4 °C (0–120 h), 25 °C (0–60 h), and 37 °C (0–30 h). The study showed that the hardness of fresh brown rice noodles was significantly increased, but the moisture content and pH were reduced with the increased storage time. The study further demonstrated that the microstructure of fresh brown rice noodles stored at 4 °C showed fewer and smaller cracks and pores. Compared to fresh brown rice noodles stored at 4 °C and 25 °C, the weakly bound water and free water of fresh brown rice noodles stored at −18 °C and 37 °C were more significantly increased and decreased, respectively (Qiao et al. 2022).

6.1 Effects of Germination of Nutritional Properties of Rice Noodles

Germination, a low-cost and high-benefit technology, has been widely applied in grain processing (Liu et al. 2022). It stimulates endogenous enzymes in brown rice, altering the functionality and structure of components, for instance, dietary fiber, protein, and starch. This process can potentially improve the edible quality and cooking of brown rice noodles (do Nascimento et al. 2022). Moreover, germination promotes the production of bioactive compounds like γ-aminobutyric acid (GABA) and polyphenols, further improving the nutritional value of the noodles. GABA, an important inhibitory neurotransmitter, is known for its sleep-improving, anti-anxiety, and blood-pressure-lowering effects. Although GABA content in brown rice is typically low and exists in small-molecule forms, germination facilitates the decarboxylation of glutamic acid, thereby increasing GABA levels (He et al. 2021).

Brown rice and its derivative products, such as brown rice noodles, have also gained popularity among consumers. Compared to polished rice, brown rice contains higher levels of non-starch nutrients, including dietary fiber, fatty acids, minerals, protein, and phenolic compounds, as these are concentrated in the bran layer. In addition to serving as an energy source, increased intake of brown rice may help lower the risk of chronic diseases (Yan et al. 2020). Rice noodles made from polished rice are renowned for their smooth and springy texture. In contrast, the high bran fiber content in brown rice poses challenges during processing and often results in lower cooking quality. Sun et al. (2023) significantly enhanced the quality of whole-grain rice noodles by incorporating extruded rice bran into refined white rice. Yi et al. (2022) found that rice bran fermentation effectively enhances the texture, improves the flavor, and increases the antioxidant activity of brown rice noodles. Similarly, Geng et al. (2021a) demonstrated that the eating quality of brown rice noodles was enhanced when brown rice was treated with ultrasound-assisted cellulase. These methods primarily involve the chemical or physical modification of bran layer fibers, decreasing the negative impact on the edible and nutritional quality of brown rice noodles. A study by Chen et al. (2024) evaluated the impact of germination on brown rice pasta including nutritional qualities, organoleptic, and texture. The study showed that germination resulted in a diminished bitter taste, reduced adhesiveness and hardness, lower cooking losses, and shorter cooking time compared to ungerminated brown rice. These changes are attributed to germination-induced alterations in the composition of brown rice. Notably, the levels of γ-aminobutyric acid, free phenolic acids, and bound phenolic acids increased by 53.43%, 21.71%, and 7.14%, respectively, while resistant starch content decreased by 21.55%, suggesting that germination emerges as a promising strategy for enhancing both the edible quality and nutritional properties of brown rice noodles (Chen et al. 2024).

6.2 Effect of Extrusion on Rice Noodle Quality

Plant-based proteins have gained increasing consumer interest as sustainable alternatives to animal protein, primarily due to their potential health benefits. Animal protein is associated with higher costs, negative environmental impacts, and limited acceptance among vegetarian and vegan consumers (Wen et al. 2019). Rice contains four primary protein fractions, namely albumin, globulin, glutenin, and prolamin. Of all rice proteins, glutelin is the most abundant, comprising 32.5% of the total protein, followed by albumin (30.9%), globulin (24.9%), and prolamin, which is a minor fraction at 11.6% (Chanput et al. 2009). The main amino acids in rice protein include arginine, cysteine, histidine, methionine, and valine. Cysteine and methionine, both sulfur-containing amino acids, play a key role in synthesizing intracellular antioxidants like *N*-acetyl cysteine and glutathione (Colovic et al. 2018). Rice contains smaller amounts of tyrosine, phenylalanine, lysine, and isoleucine. However, the lysine (3.8%) and leucine (8.2%) content in rice are higher compared to wheat (2.8% and 6.7%, respectively) (Singh et al. 2016). Furthermore, rice protein exhibits superior nutritional quality, true digestibility, and biological value compared to soy protein isolate (Han et al. 2015) and other cereals (Eggum 1979). Due to its high lysine content and high-quality amino acid profile, rice protein is considered hypoallergenic, making it suitable for human consumption (Wanasundara et al. 2024). Rice protein is recommended as a dietary alternative for infants and the elderly (Alves et al. 2022). Studies have demonstrated its functional properties, including antioxidant activity (Li et al. 2021), anti-hypertensive (Zhang et al. 2023), anti-cancer (Wang et al. 2023), and anti-obesity effects (Sivamaruthi et al. 2023). Rice flour is commonly used as a key ingredient in gluten-free products due to its neutral flavor, easy digestibility, and hypoallergenic properties. However, the absence of gluten in gluten-free spaghetti leads to quality issues, such as increased cooking loss, higher stickiness, and reduced firmness (Ahmed et al. 2024).

Various techniques have been demonstrated to improve the process of noodle making. Extrusion is a widely used food processing technology in the food industry. This method combines mixing, shearing, kneading, cooking, and compressing to process food materials. Extrusion also denatures proteins, solubilizes fibers, gelatinizes starch, and facilitates the cross-linking of biopolymers. It is extensively applied in the production of cereal products, snacks, pasta, and other food items (Leonard et al. 2020). Currently, extrusion technology is primarily categorized into single-screw and twin-screw extrusion. Research indicates that key parameters such as screw speed, moisture content, and temperature in the processing zones significantly affect the quality of extruded products (Ribeiro et al. 2024). The extrusion cooking method is a promising alternative technology for producing gluten-free pasta-like products. Within the extruder, starch granules are disrupted, starch undergoes gelatinization, and proteins are denatured, enabling the formation of gluten-free pasta. The extrusion conditions, which influence starch gelatinization and protein denaturation, can also induce intramolecular and intermolecular interactions between these polymers, further modifying the product's structure and properties

(Yu et al. 2023). A study by Detchewa et al. (2022) evaluated the cooking quality and physicochemical properties of gluten-free rice spaghetti by substituting rice flour with rice protein concentrate at 0% (control), 2.5%, 5.0%, 7.5%, and 10% (w/w) and processed using a single-screw extruder. The data showed that substituting rice flour with 5%–10% rice protein significantly reduced cooking time and cooking loss. This result aligns with the findings of Kim et al. (2014), who reported that supplementing rice noodles with 10% rice protein isolate decreased cooking loss from 11.83% to 6.69%. Gluten-free rice spaghetti containing 5%–10% rice protein exhibited a more porous structure compared to gluten-free rice spaghetti made without rice protein (Detchewa et al. 2022).

Soybean protein is regarded as a key component of plant-based dietary patterns due to its high quality and health benefits. Specifically, soybean 7S globulin (β-conglycinin) is known for its digestion resistance (Li et al. 2023), making it an ideal ingredient for low glycemic index (GI) foods. The data showed that increasing the addition of soybean 7S protein reduced starch digestibility, leading to a decrease in the estimated glycemic index (eGI) and glycemic load (GL) (Qiao et al. 2023). However, higher levels of 7S protein addition resulted in a gradual increase in cooking loss and breakage rates of rice noodles. Water absorption initially increased but decreased as the 7S protein content increased. The sensory evaluation showed that the taste of rice noodles remained acceptable when the 7S protein addition was kept within 9% (Qiao et al. 2023). The interaction between starch and soybean 7S protein is primarily mediated by weak forces, including hydrophobic interactions and hydrogen bonds. Intriguingly, Tian et al. (2022) found that extrusion cooking of rice bran prevented the reduction in moisture content and pH, decreased the cooking loss rate, and increased the percentage of free and weakly bound water in fresh brown rice noodles during storage at −18 °C, 4 °C, 25 °C, and 37 °C. The microstructure of the fresh brown rice noodles prepared from extruded rice bran exhibited a well-formed, regular porous structure. Differential scanning calorimetry and X-ray diffraction analysis showed that starch retrogradation in fresh brown rice noodles prepared from extruded rice bran was inhibited or delayed during storage at various temperatures, particularly when stored at 4 °C. The data suggested that the storage quality of fresh brown rice noodles made with rice bran through extrusion cooking could be improved, with starch retrogradation in fresh brown rice noodles prepared from extruded rice bran being inhibited (Tian et al. 2022).

6.3 Effects of Functional Ingredients on Nutritional Properties of Rice Noodles

Brown rice noodles are often characterized by poor eating quality and slow digestion. Several methods such as stabilizing rice bran through microwave treatment and ultrafine grinding, have been used to improve the eating quality of brown rice noodles (Geng et al. 2021b). The slow digestion of brown rice noodles is evident in their lower starch digestibility compared to polished rice noodles. However, the predicted glycemic index (pGI) of brown rice noodles is around 70, which remains relatively high and does not fully meet the health needs of consumers (Geng et al. 2021a). A previous study has shown that red lentil protein is more effective than rice protein in reducing the starch digestibility of brown rice noodles (Geng et al. 2021b). The previous study has demonstrated that the thermostability and anti-retrogradation ability of brown rice flour are significantly reduced after adding untreated red lentil protein, whereas the elastic and viscous moduli are increased. Moreover, the cooking loss rate of brown rice noodles rose from 4.41% to 5.68%. However, when red lentil protein was treated with ultrasound and enzymes, the protein-starch interaction was enhanced. Furthermore, rice noodles made with ultrasound-assisted enzyme-treated red lentil protein exhibited lower starch digestibility than brown rice noodles, with significantly higher protein digestibility compared to rice noodles containing either ultrasound- or enzyme-treated protein, suggesting that the addition of ultrasound-assisted enzyme-treated red lentil protein may enhance nutritional value and produce high-quality brown rice noodles (Lin et al. 2022).

A study by Chen et al. (2022) evaluated the effects of partial replacement (0%–20%, w/w) of rice flour with potato starch on the *in vitro* digestibility, texture, microstructure, and physicochemical properties of instant rice noodles. The study demonstrated that the incorporation of rice noodles with 5%–15% potato starch (potato starch and *indica* rice flour mixed at ratios of 5:95, 10:90, 15:85, and 20:80, respectively) showed modified textural and appearance properties. The mutually exclusive phosphate groups in potato starch facilitated the expansion of starch granules, and thus resulting in the formation of a more compact microstructure and higher-ordered crystalline structures. These changes enhanced the digestive and cooking characteristics of the noodles, contributing to a noticeable increase in slowly digestible starch content (26.09%–31.34%), suggesting that blending potato starch is both a cost-efficient and effective strategy to improving the textural and nutritional properties of instant rice noodles. In another study, Sundaresan et al. (2023) evaluated the functional characteristics of the defatted coconut flour on the antioxidant properties, textural, cooking qualities, color, and nutritional composition in rice noodles. The addition of defatted coconut flour significantly enhanced oil absorption capacity, water absorption capacity, and water solubility index. Nonetheless, the viscosity, emulsion activity, foaming capacity, and bulk density are significantly decreased in rice noodles. Nutrition analysis indicated that defatted coconut flour-enriched formulations had significantly higher levels of minerals, flavonoids, polyphenols, antioxidants, proteins, and crude fiber compared to the control. Moreover, the addition of defatted coconut flour significantly improved noodle thickness, cooking time, cooking loss, water uptake ratio, and swelling volume. Incorporating defatted coconut flour also led to a decrease in gumminess, resilience, chewiness, cohesiveness, springiness, and hardness of rice noodles (Sundaresan et al. 2023). The data implied that noodles made with defatted coconut flour contained higher levels of fiber and protein compared to those made with rice flour.

6.4 Effects of Fermentation on Nutritional Properties of Rice Noodles

A study by Yi et al. (2022) evaluated the effects of rice bran fermented with *Lactobacillus plantarum* on antioxidant activity, volatile profiles, and palatability of brown rice noodles. The data demonstrated that bran fermentation decreased the cooking loss of bran-fermented brown rice noodles while enhancing the sensory and elongation attributes. Fermentation significantly decreased aldehyde levels and increased alcohol levels, contributing to more pleasant flavors in bran-fermented brown rice noodles. Furthermore, fermentation enhanced free phenolics while reducing bound phenolics (Yi et al. 2022). Moreover, the ferric-reducing antioxidant power and DPPH radical scavenging activity of free phenolics in bran fermented brown rice noodles significantly increased compared to unfermented brown rice noodles. This study demonstrated that rice bran fermentation is an effective method for improving the palatability, enhancing the flavor, and increasing the antioxidant activity of brown rice noodles (Yi et al. 2022). These findings may contribute to advancing processing technologies for brown rice noodles.

References

Ahmed MW, Jothi JS, Saifullah M et al (2024) Chapter 4 – Impact of drying temperature on textural, cooking quality, and microstructure of gluten-free pasta. In: Gull A, Nayik GA, Brennan C (eds) Development of gluten-free pasta. Academic Press, pp 65–110

Alves JB, Rodrigues MHP, Duarte FA et al (2022) Rice bran and its potential to complement the nutritional needs of children and elderly. Plant Foods Hum Nutr 78:86–92

Bhattachrya M, Zee SY, Corke H (1999) Physicochemical properties related to quality of rice noodles. Cereal Chem 76:861–867

Cáceres PJ, Peñas E, Martínez-Villaluenga C et al (2019) Development of a multifunctional yogurt-like product from germinated brown rice. LWT Food Sci Technol 99:306–312

Chanput W, Theerakulkait C, Nakai S (2009) Antioxidant properties of partially purified barley hordein, rice bran protein fractions and their hydrolysate. J Cereal Sci 49:422–428

Chen J, Yang S, Zhang M et al (2022) Effects of potato starch on the characteristics, microstructures and quality attributes of indica rice flour and instant rice noodles. Int J Food Sci Technol 57:2285–2297

Chen R, Yan X, Cai M et al (2024) Impact of germination on the edible quality and nutritional properties of brown rice noodles. Foods 13:2152

Colovic MB, Vasic VM, Djuric DM et al (2018) Sulphur-containing amino acids: protective role against free radicals and heavy metals. Curr Med Chem 25:324–335

Detchewa P, Prasajak P, Phungamngoen C et al (2022) Substitution of rice flour with rice protein improved quality of gluten-free rice spaghetti processed using single screw extrusion. LWT 153:112512

do Nascimento LÁ, Abhilasha A, Singh J et al (2022) Rice germination and its impact on technological and nutritional properties: a review. Rice Sci 29:201–215

Eggum BO (1979) The nutritional value of rice in comparison with other cereals. In: Proceedings of the workshop on chemical aspects of rice grain quality. International Rice Research Institute, Los Banos, pp 91–111

Geng D-H, Zhou S, Wang L et al (2020) Effects of slight milling combined with cellulase enzymatic treatment on the textural and nutritional properties of brown rice noodles. LWT Food Sci Technol 128:109520

Geng D, Lin Z, Liu L et al (2021a) Effects of ultrasound-assisted cellulase enzymatic treatment on the textural properties and in vitro starch digestibility of brown rice noodles. LWT Food Sci Technol 146:111543

Geng D, Liu L, Lin Z et al (2021b) Effects of red lentil protein addition on textural quality and starch digestibility of brown rice noodles. Int J Food Sci Technol 56:6656–6666

Guiné RPF, Florença SG, Barroca MJ et al (2020) The link between the consumer and the innovations in food product development. Foods 9:1317

Han S-K, Chee KM, Cho S-J (2015) Nutritional quality of rice protein in comparison to animal and vegetable protein. Food Chem 172:766–769

He LY, Yang Y, Ren LK et al (2021) Effects of germination time on the structural, physicochemical and functional properties of brown rice. Int J Food Sci Technol 57:1902–1910

Huang M, Xiao Z, Chen J et al (2021) Yield and quality of brown rice noodles processed from early-season rice grains. Sci Rep 11:18668

Jiao A, Yang Y, Li Y et al (2020) Structural properties of rice flour as affected by the addition of pea starch and its effects on textural properties of extruded rice noodles. Int J Food Prop 23:809–819

Kim Y, Kee JI, Lee S et al (2014) Quality improvement of rice restructured with rice protein isolate and transglutaminase. Food Chem 145:409–416

Leonard W, Zhang P, Ying D et al (2020) Application of extrusion technology in plant food processing byproducts: an overview. Compr Rev Food Sci Food Saf 19:218–246

Li Y, Liang JF, Yang MY et al (2015) Traditional chinese rice noodles: history, classification and processing methods. Cereal Food World 60:123–127

Li F, Wu X, Wu W (2021) Effects of oxidative modification by malondialdehyde on the in vitro digestion properties of rice bran protein. J Cereal Sci 97:103158

Li T, Han K, Feng G et al (2023) Bile acid profile influences digestion resistance and antigenicity of soybean 7S protein. J Agric Food Chem 71:2999–3009

Lin Z, Liu L, Qin W et al (2022) Changes in the quality and in vitro digestibility of brown rice noodles with the addition of ultrasound-assisted enzyme-treated red lentil protein. Int J Food Sci Technol 57:1150–1160

Liu S, Wang W, Lu H et al (2022) New perspectives on physiological, biochemical and bioactive components during germination of edible seeds: a review. Trends Food Sci Technol 123:187–197

Muniandy SD, Gannasin SP (2019) Physicochemical and sensory characteristics of brown rice (*Oryza sativa*) noodles substituted with mung bean (*Vigna radiata*) powder. Malays J Nutr 25:351–359

Park J, Kim H-S (2023) Rice-based gluten-free foods and technologies: a review. Foods 12:4110

Qiao C-C, Tian X-H, Wang L-X et al (2022) Quality characteristics, texture properties and moisture migration of fresh brown rice noodles under different storage and temperatures conditions. J Cereal Sci 104:103434

Qiao F, Wang L, Guan C et al (2023) Effects of soybean 7S protein on the quality and digestibility of dry rice noodles under twin-screw extrusion process. Int J Food Sci Technol 58:463–472

Ribeiro G, Piñero M-Y, Parle F et al (2024) Optimizing screw speed and barrel temperature for textural and nutritional improvement of soy-based high-moisture extrudates. Foods 13:1748

Saleh M, Abuwaar Z, Akash MW et al (2014) Effect of stabilized rice bran fractions on the formation of rice flour pasting properties. Cereal Chem 91:603–609

Singh AK, Singh R, Subramani R et al (2016) Current genomics, molecular approaches to understand nutritional potential of coarse cereals. Curr Genomics 17:1–16

Sivamaruthi BS, Alagarsamy K, Thangaleela S et al (2023) Composition, microbiota, mechanisms, and anti-obesity properties of rice bran. Foods 12:1300

Sun Z, Lyu Q, Zhuang K et al (2023) Impact of different preparation methods on the properties of brown rice flour and the cooking stability of brown rice noodles and the underlying mechanism: microstructure, starch-protein distribution, moisture migration. LWT Food Sci Technol 181:114697

Sundaresan T, Subramaniam S, Chinnapa V et al (2023) Effect of defatted coconut flour on functional, nutritional, textural and sensory attributes of rice noodles. Int J Food Sci Technol 58:5077–5088

Tian X-H, Tan B, Wang L-X et al (2022) Effect of rice bran with extrusion cooking on quality and starch retrogradation of fresh brown rice noodles during storage at different temperatures. Cereal Chem 99:1296–1307

Wanasundara JPD, Kumagai H, Kasch AP et al (2024) Chapter 3 – Rice: a source of plant protein with many valuable processing co-products. In: Nadathur S, Wanasundara JPD, Scanlin L (eds) Sustainable protein sources: advances for a healthier tomorrow. Academic Press, pp 55–75

Wang N, Cui X, Duan Y et al (2023) Potential health benefits and food applications of rice bran protein: research advances and challenges. Food Rev Int 39:3578–3601

Wen C, Zhang J, Yao H et al (2019) Advances in renewable plant-derived protein source: the structure, physicochemical properties affected by ultrasonication. Ultrason Sonochem 53:83–98

Yan XD, Liu CM, Huang A et al (2020) The nutritional components and physicochemical properties of brown rice flour ground by a novel low temperature impact mill. J Cereal Sci 92:102927

Yi C, Zhu H, Zhang Y et al (2021) The role of indica starch in the mechanism of formation of fresh rice noodles. J Cereal Sci 99:103212

Yi C, Xie L, Cao Z et al (2022) Effects of rice bran fermented with *Lactobacillus plantarum* on palatability, volatile profiles, and antioxidant activity of brown rice noodles. Int J Food Sci Technol 57:5048–5056

Yu X, Wang L, Zhang J et al (2023) Understanding effects of glutelin on physicochemical and structural properties of extruded starch and the underlying mechanism. Carbohydr Polym 304:120513

Zhang L, Miao J, Guo J et al (2023) Two novel angiotensin I-converting enzyme (ACE) inhibitory peptides from rice (*Oryza sativa* L.) bran protein. J Agric Food Chem 71:4153–4162

Chapter 7
Glycemic Index of the Rice Noodles

Contents

Rice noodles, a traditional staple food in Southern China and Southeast Asia, have a history spanning over 2300 years (Luo et al. 2024). Variations in manufacturing and cooking techniques have led to the development of distinct regional cuisines, including Guangxi, Yunnan, and Changde rice noodles. Despite regional differences, the smooth and sinewy texture of rice noodles contributes to their widespread popularity. However, rice noodles are associated with a high GI and low dietary fiber content due to their predominant ingredient, rice starch (Luo et al. 2024). These properties make them unsuitable for individuals with diabetes or hyperglycemia (Luo et al. 2024). Even in healthy individuals, long-term consumption of high-GI staples like rice noodles can increase the risk of obesity and hyperglycemia (Tan et al. 2018). To address these concerns, there is an urgent need to develop low-GI rice noodles tailored for specific populations, including those with diabetes, hyperglycemia, and obesity, seeking hypoglycemic staple food options.

B. L. Tan, M. E. Norhaizan, *Rice Noodles: Quality Improvement, Processing Techniques and Nutritional Aspects*, SpringerBriefs in Food, Health, and Nutrition,
https://doi.org/10.1007/978-3-031-98649-9_7

7.1 Glycemic Index

Nutrition plays a pivotal role in managing type 2 diabetes by significantly influencing weight and metabolic control. Medical nutrition therapy provides an evidence-based framework for managing diabetes through lifestyle modifications. These strategies are strongly recommended as the first line of intervention, even before initiating pharmaceutical treatments. However, many physicians often bypass this approach, despite evidence showing that anti-diabetic medications alone are insufficient to halt disease progression. Nearly half of type 2 diabetes patients are expected to require insulin therapy within 10 years of diagnosis (Swinnen et al. 2009). Therefore, it is essential to develop and implement effective dietary strategies to control hyperglycemia and slow the progression of type 2 diabetes through targeted diet and lifestyle changes. Adopting appropriate staple food modifications is a practical approach to diabetes management. The primary goal of dietary interventions is to stabilize blood glucose levels by incorporating low-GI foods into the diet (Gerontiti et al. 2024).

The quality and quantity of carbohydrates are widely recognized as key determinants of glycemic response (Tan and Norhaizan 2024b). Dietary patterns emphasizing low glycemic index and/or low glycemic load have been shown to enhance glycemic control in individuals with type 2 diabetes. The glycemic index of a carbohydrate-rich food estimates how quickly its carbohydrates are digested and absorbed, impacting blood glucose levels (Wee and Henry 2020). The glycemic index (GI) is also defined as the ratio of the incremental area under the glucose response curve (AUC) for a test food containing a standard amount of glucose to that of a reference food (typically glucose or white bread) (Romão et al. 2021). High glycemic index foods, which are rapidly digested and absorbed, leading a sharp increase in blood glucose and insulin levels. In contrast, low glycemic index foods are digested more slowly, resulting in gradual and smaller postprandial glucose and insulin responses. High glycemic index foods have been associated with pathological conditions, including an increased risk of obesity, even in non-diabetic individuals (Pereira et al. 2015). The GI of starchy foods can be modified through processing methods or by combining them with other foods. Additionally, the GI of food also depends on the amount and type of carbohydrates it contains, as these factors influence insulin secretion and metabolism (Lal et al. 2021).

Several factors influence a food's glycemic index (Obadi and Xu 2024) including the type of carbohydrate (such as amylopectin-rich starch is digested faster than amylose-rich starch), the content of protein, fat, and fiber (soluble fiber slows postprandial glycemia, while insoluble fiber does not), and food characteristics such as particle size and pH (lower pH foods delay glucose absorption). The GI is determined *in vivo* by measuring the area under the curve (AUC) of postprandial glucose levels following ingestion of a carbohydrate-rich food containing 50 g of digestible carbohydrate, compared to the AUC after consuming 50 g of glucose (Brand-Miller et al. 2003). Foods and food products are categorized based on their glycemic index (GI) values, which reflect their ability to raise postprandial blood glucose levels.

High-GI foods have a value greater than 70, medium-GI foods range from 56 to 69, and low-GI foods are less than 55. This classification assumes that the GI of the standard reference food, either white bread or glucose, is set at 100 (Brand-Miller et al. 2003).

7.2 Factors Affecting Glycemic Index of Rice Noodles

Several studies have reported that increasing dietary fiber or resistant starch is positively correlated with higher gelatinization temperatures (Nagy et al. 2021). Similarly, an increase in amylose content raises gelatinization temperatures while reducing enthalpies (Zhao et al. 2022). Waxy rice, characterized by low amylose but high amylopectin content, exhibits higher gelatinization enthalpy due to its higher crystallinity and lower amorphous region content (Fei et al. 2023). Factors such as genetic origin, environmental conditions, and the age of the parent plant influence the gelatinization temperatures of starchy samples (Rahim et al. 2020; Kim and Kim 2021). The mixtures with high amylose and resistant starch content exhibited lower gelatinization enthalpies, indicating they were easier to cook. This characteristic could be advantageous for industrial applications by reducing energy consumption and processing time (Srikaeo and Sangkhiaw 2014).

A study showed that an addition of high-amylose maize starch demonstrated its potential to lower the glycemic index of rice noodle products (Srikaeo and Sangkhiaw 2014). Rice and most rice-based foods are known for their high glycemic index, posing a challenge for the food industry to develop low glycemic index rice products. As rice is a staple food for individuals with celiac disease, creating low glycemic index rice products could offer additional health benefits. Low glycemic index foods are particularly valuable as they reduce insulin response and promote greater utilization of stored fat. Furthermore, resistant starch has been studied for its potential health benefits, making high-resistant starch and low glycemic index foods beneficial for managing obesity and diabetes (Guo et al. 2022).

The previous study has demonstrated that starch digestibility can be modulated by adjusting amylose and resistant starch levels in the ingredients. Foods rich in amylose have been associated with lower blood glucose levels and slower gastrointestinal emptying compared to those with low amylose content (Li et al. 2024). Amylose content significantly impacts GI values and resistant starch levels, as evidenced by studies showing that increasing amylose content increases resistant starch content in various food systems (Li et al. 2022).

Rice noodles produced from mixtures containing 60 g of maize starch per 100 g of dry sample exhibited glycemic index values below 55 (as determined by the in vivo method). Therefore, these noodles can be classified as low-GI foods (Srikaeo and Sangkhiaw 2014). Glycemic index values are influenced by various factors, including mass transfer resistance, degree of branching (affecting steric hindrance), molecular structure, the ratio of amylose to amylopectin, and starch granule morphology (Subaitha et al. 2024). Additionally, other components in rice flour can impact the rate of starch digestion and GI (Kunyanee and Luangsakul 2022).

Texture is a critical consideration for the food industry when producing starchy foods. In Thailand, rice noodles with high tensile strength and extensibility are preferred, although consumer preferences may vary in other regions. Furthermore, other components in the mixtures, such as proteins from rice flour, can also affect the texture of the final products (Hager et al. 2012). Rice with high amylose content typically produces fluffy, dry textures, while low-amylose rice results in sticky, chewy, and moist textures after cooking. The amylose-to-amylopectin ratio affects the hardness of rice starch gels, with high-amylose varieties generally exhibiting greater hardness (Varghese et al. 2022). Hardness is often negatively correlated with adhesiveness, and thus with amylose content. As a result, rice noodles made from high-amylose mixtures tend to have greater hardness and a fluffy texture but exhibit lower tensile strength and break distance (extensibility) (Srikaeo and Sangkhiaw 2014).

7.3 Starch Digestibility and Glycemic Index

Starch is the primary component of carbohydrate-rich food crops, which also contain health-promoting compounds such as resistant starch (RS) and polysaccharide fibers. During digestion, carbohydrates are broken down into various sugars, but only glucose directly impacts blood sugar levels, triggering insulin release. This process not only regulates blood glucose but also affects satiety and supports gut health by exerting prebiotic effects in the large intestine (Liu et al. 2020).

Starch is composed of two types of polyglucans: amylose and amylopectin. Amylose is a mostly linear polysaccharide of D-glucose units linked by $\alpha(1 \rightarrow 4)$ glycosidic bonds, with occasional branch points. In contrast, amylopectin is a larger, highly branched molecule, containing both $\alpha(1 \rightarrow 4)$ glycosidic bonds and $\alpha(1 \rightarrow 6)$ branch points (Dong et al. 2025). Amylose has long, unbranched chains, while amylopectin features numerous short-chain branches, contributing to its high molecular weight.

In cereals, starch typically comprises 18–33% amylose and 70–80% amylopectin, with an approximate ratio of 30:70 (Wang et al. 2024b). Unpolished (brown) rice contains about 72–82% of starch, while polished (milled) rice contains ~90% starch. Rice, maize, and barley cultivars with less than 1% amylose are classified as waxy varieties. The amylose-to-amylopectin ratio influences the texture and digestibility of starchy grains. For example, higher amylopectin content contributes to the floury texture of tubers, such as mature potatoes. The increased branching in amylopectin enhances its surface area, facilitating access to starch-hydrolyzing enzymes (Li et al. 2023).

The rate-limiting enzyme in starch synthesis is adenosine 5′-diphosphate (ADP) pyrophosphorylase (AGPase), which catalyzes the production of starch using ADP-glucose as a substrate (Ahmad et al. 2024). Starch synthases extend the chain length of the molecule, while granule-bound starch synthase I (GBSS I) is specifically required for amylose synthesis, in coordination with certain isoforms of debranching enzymes. Amylopectin biosynthesis involves multiple isoforms of starch synthases, starch branching enzymes (SBE), and starch debranching enzymes (SDBE). The overall synthesis of starch is primarily regulated by the activity of GBSSI and various isoforms of starch synthases. Previous studies have demonstrated that starchy foods elevate postprandial blood glucose and insulin levels in both healthy and diabetic individuals (Oba-Yamamoto et al. 2022). However, these responses are influenced by the rate at which starch is digested in the human body.

7.4 Amylose, Resistant Starch, and Glycemic Index of Rice Noodles

Rice is a staple food in Asia and is one of the vitally important cereal crops. Nonetheless, it has been demonstrated to have a relatively high glycemic index than other starchy foods. Several studies revealed that the glycemic index of rice is between 54 to 121 (Jenkins et al. 1981, 1984; Brand et al. 1985; Hu et al. 2004; Jaisut et al. 2008, 2009; Shobana et al. 2012).

There are several factors that affects the metabolic and physicochemical properties of rice. Among these factors, amylose content is one of the crucial factors that are usually used to predict the insulin responses, blood glucose, and starch digestion rate of rice. In particular, starchy foods that are rich in amylose are linked to the slower emptying of the human gastrointestinal tract and lower blood glucose levels than low levels of amylose (Frei et al. 2003). High-amylose rice showed lower glycemic index levels compared to low-amylose varieties (Denardin et al. 2012). However, some studies found that rice with similar amylose content may differ in glycemic response and starch digestibility (Panlasigui et al. 1991). In addition to amylose/amylopectin ratios, starch properties, for instance, non-starch components, degree of polymerization, surface pores and channels, degree of crystallinity, crystalline pattern, architecture, and granule size (Parada and Aguilera 2012).

Besides amylose, resistant starch has received great interest for its functional properties and health benefits (Bojarczuk et al. 2022). Resistant starch positively affects the digestive tract, the glycemic index, the blood cholesterol level, the microbial flora, and modulates diabetes (Chen et al. 2024). The starch digestion rate in foods, and consequently their glycemic index can be modified by several techniques. These may involve the alteration of functional ingredients, using less degree of polishing in milled rice, extrusion, applying processing techniques, formation of starch lipid complexes, and using low or no-calorie sugars (Srikaeo and Sangkhiaw 2014).

Rice noodles are a potential rice product that exerts a lower glycemic index value compared to other rice products. In general, rice noodles are made of high amylose rice to improve noodle textures (Liu et al. 2021). The processing techniques applied tempering of steamed rice sheets and steaming of rice flour slurry. Subsequently, this may increase resistant starch or slowly digestible starch fraction and thereby leading to a reduction of glycemic index through starch retrogradation (Zhang et al. 2011) and hydrothermal treatment (Guha et al. 2011).

The previous study showed that the incorporation of high-amylose maize starch altered the properties of rice noodle mixtures and the resulting rice noodle products. This starch modified the resistant starch and amylose content in the mixtures. An increase in amylose and resistant starch content significantly influenced the mixture's properties. Differential scanning calorimetry analysis showed minimal changes in gelatinization temperatures, a slight increase in peak temperature and higher amylose levels. Nonetheless, the gelatinization enthalpy decreased as the proportion of maize starch in the mixtures increased (Srikaeo and Sangkhiaw 2014). A recent study found that the incorporation of transglutaminase, soybean protein isolate, and konjac glucomannan into *indica* rice flour enhances rice noodle quality and reduces glycemic impact (Wang et al. 2024a).

7.5 Health Implications of Glycemic Index in Rice Noodles

7.5.1 Diabetes Mellitus

Diabetes and hyperglycemia are increasingly prevalent chronic metabolic disorders with significant global health implications. Diabetes mellitus is broadly categorized into two types: type 1 and type 2. Type 1 diabetes, or insulin-dependent diabetes, results from an absolute deficiency in insulin secretion caused by autoimmune destruction of pancreatic β-cells (Tatovic et al. 2023). In contrast, type 2 diabetes, a chronic metabolic condition, is characterized by a combination of decreased insulin production and insulin resistance. This leads to elevated blood glucose levels due to the body's reduced ability to utilize insulin effectively. Unlike type 1 diabetes, which is caused by an inability to produce insulin, type 2 diabetes arises from a progressive decline in insulin sensitivity and pancreatic function. Historically referred to as adult-onset diabetes, type 2 diabetes is increasingly diagnosed in younger populations due to rising obesity rates, sedentary lifestyles, and genetic predisposition (Goyal et al. 2023).

Diabetes manifests through a range of clinical symptoms, including frequent urination, excessive hunger, unexplained weight loss, breathing difficulties, fatigue, and pain (Hajam et al. 2022). Additionally, diabetes is closely associated with numerous complications, particularly cardiovascular disease (Tan et al. 2018; Tan and Norhaizan 2024a). Globally, diabetes affects approximately 422 million people, with the majority residing in low- and middle-income countries. The disease is directly responsible for 1.5 million deaths each year. Over recent decades, both the incidence and the number of diabetes cases have steadily risen, indicating a significant public health threat. This increase is largely attributed to factors such as rapid urbanization, dietary changes that elevate the consumption of high-calorie foods, and a decline in physical activity. Given the growing prevalence of diabetes, there is an urgent need for comprehensive public health interventions to mitigate the impact of type 2 diabetes, particularly in countries like Malaysia (World Health Organization 2024).

Oxidative stress plays a central role in the onset and progression of type 2 diabetes. It is characterized by an imbalance between the body's antioxidant defenses and the production of reactive oxygen species (ROS) (Caturano et al. 2023). Chronic hyperglycemia, a hallmark of type 2 diabetes, contributes significantly to the overproduction of ROS. These highly reactive molecules can cause severe damage to cellular components, including DNA, proteins, and lipids, thereby exacerbating the progression of the disease (Zhang et al. 2021).

Oxidative stress contributes significantly to type 2 diabetes through the degradation of the insulin signaling pathway. ROS can modify key signaling molecules, disrupting the normal cellular response to insulin and worsening insulin resistance (Bhatti et al. 2022). Additionally, oxidative stress can impair the function of pancreatic beta cells, reducing insulin secretion and further aggravating hyperglycemia. Elevated blood glucose levels enhance oxidative stress, which in turn exacerbates insulin resistance and beta-cell dysfunction, creating a vicious cycle that accelerates the progression of type 2 diabetes (Eguchi et al. 2021).

Oxidative stress is closely linked to the development of several complications associated with diabetes, including retinopathy, neuropathy, nephropathy, and cardiovascular diseases (Caturano et al. 2023). ROS contributes to cardiovascular risks by damaging endothelial cells, promoting inflammation, and accelerating the formation of atherosclerotic plaques within the vascular system. In peripheral tissues, oxidative stress can lead to damage to the kidneys and nerves, contributing to neuropathy and nephropathy (Scioli et al. 2020). These complications highlight the critical role of oxidative stress in the pathogenesis of diabetes-related disorders.

The increasing prevalence of type 2 diabetes worldwide is driven by a combination of genetic and lifestyle factors. Given its significant role in the development of insulin resistance and diabetes-related complications, targeting oxidative stress is crucial in treatment strategies (Tan and Norhaizan 2024a). Approaches such as improving glycemic control, utilizing antioxidant therapies, and implementing lifestyle modifications can reduce oxidative stress, potentially mitigating the long-term consequences of type 2 diabetes. These interventions may enhance the quality of life for individuals living with this chronic condition.

References

Ahmad D, Ying Y, Bao J et al (2024) Understanding starch biosynthesis in potatoes for metabolic engineering to improve starch quality: a detailed review. Carbohydr Polym 346:122592

Bhatti JS, Sehrawat A, Mishra J et al (2022) Oxidative stress in the pathophysiology of type 2 diabetes and related complications: current therapeutics strategies and future perspectives. Free Radic Biol Med 184:114–134

Bojarczuk A, Skąpska S, Khaneghah AM et al (2022) Health benefits of resistant starch: a review of the literature. J Funct Foods 93:105094

Brand JC, Nicholson PL, Thorburn AW et al (1985) Food processing and the glycemic index. Am J Clin Nutr 42:1192–1196

Brand-Miller J, Hayne S, Petocz P et al (2003) Low-glycemic index diets in the management of diabetes: a meta-analysis of randomized controlled trials. Diabetes Care 26:2261–2267

Caturano A, D'Angelo M, Mormone A et al (2023) Oxidative stress in type 2 diabetes: impacts from pathogenesis to lifestyle modifications. Curr Issues Mol Biol 45:6651–6666

Chen Z, Liang N, Zhang H et al (2024) Resistant starch and the gut microbiome: exploring beneficial interactions and dietary impacts. Food Chem X 21:101118

Denardin CC, Boufleur N, Reckziegel P et al (2012) Amylose content in rice (*Oryza sativa*) affects performance, glycemic and lipid metabolism in rats. Ciência Rural 42:381–387

Dong Y, Li Z, Kong H et al (2025) Correlation analysis of starch molecular structure and film properties via rearrangements of glycosidic linkages by 1,4-α-glucan branching enzyme. Carbohydr Polym 348:122908

Eguchi N, Vaziri ND, Dafoe DC, Ichii H (2021) The role of oxidative stress in pancreatic β cell dysfunction in diabetes. Int J Mol Sci 22(4):1509

Fei Y, Han H, Yu B et al (2023) Physicochemical properties of waxy rice varieties with contrasting hardening rates. Int J Food Prop 26:775–1786

Frei M, Siddhuraju P, Becker K (2003) Studies on the *in vitro* starch digestibility and the glycemic index of six different indigenous rice cultivars from the Philippines. Food Chem 83:395–402

Gerontiti E, Shalit A, Stefanaki K et al (2024) The role of low glycemic index and load diets in medical nutrition therapy for type 2 diabetes: an update. Hormones 23:655–665

Goyal R, Jialal I, Singhal M (2023) Type 2 diabetes. National Center for Biotechnology Information; StatPearls Publishing

Guha M, Umesh SS, Reddy SY et al (2011) Functional properties of slow carbohydrate digestible rice produced adapting hydrothermal treatment. Int J Food Prop 14:1305–1317

Guo J, Tan L, Kong L (2022) Multiple levels of health benefits from resistant starch. J Agric Food Res 10:100380

Hager AS, Wolter A, Czerny M et al (2012) Investigation of product quality, sensory profile and ultrastructure of breads made from a range of commercial gluten-free flours compared to their wheat counterparts. Eur Food Res Technol 235:333–344

Hajam YA, Rani R, Malik JA et al (2022) Diabetes mellitus: signs and symptoms, epidemiology, current prevention, management therapies, and treatments. In: Antidiabetic potential of plants in the era of omics, 1st edn. Apple Academic Press, p 47

Hu P, Zhao H, Duan Z et al (2004) Starch digestibility and the estimated glycemic score of different types of rice differing in amylose contents. J Cereal Sci 40:231–237

Jaisut D, Prachayawarakorn S, Varanyanond W et al (2008) Effects of drying temperature and tempering time on starch digestibility of brown fragrant rice. J Food Eng 86:251–258

Jaisut D, Prachayawarakorn S, Varanyanond W et al (2009) Accelerated aging of jasmine brown rice by high-temperature fluidization technique. Food Res Int 42:674–681

Jenkins DJA, Wolever TMS, Taylor RH (1981) Glycemic index of foods: a physiological basis for carbohydrate exchange. Am J Clin Nutr 34:362–366

Jenkins DJA, Wolever TMS, Jenkins AL et al (1984) The glycemic response to carbohydrate foods. Lancet 18:388–391

Kim K-H, Kim J-Y (2021) Understanding wheat starch metabolism in properties, environmental stress condition, and molecular approaches for value-added utilization. Plants 10:2282

Kunyanee K, Luangsakul N (2022) The impact of heat moisture treatment on the physicochemical properties and *in vitro* glycemic index of rice flour with different amylose contents and associated effects on rice dumpling quality. LWT 154:112694

Lal MK, Singh B, Sharma S et al (2021) Glycemic index of starchy crops and factors affecting its digestibility: a review. Trends Food Sci Technol 111:741–755

Li C, Dhital S, Gidley MJ (2022) High-amylose wheat bread with reduced *in vitro* digestion rate and enhanced resistant starch content. Food Hydrocoll 123:107181

Li M, Daygon VD, Solah V et al (2023) Starch granule size: does it matter? Crit Rev Food Sci Nutr 63:3683–3703

Li J, Yoshimura K, Sasaki M et al (2024) The consumption of high-amylose rice and its effect on postprandial blood glucose levels: a literature review. Nutrients 16:4013

Liu H, Zhang M, Ma Q et al (2020) Health beneficial effects of resistant starch on diabetes and obesity via regulation of gut microbiota: a review. Food Funct 11:5749–5767

Liu Y, Chen X, Xu Z et al (2021) Addition of waxy, low- or high-amylose rice starch differentially affects microstructure, water migration, texture and cooking quality of dried potato starch noodles. Int J Food Sci Technol 56:5619–5628

Luo X, Chang X, Zhang K et al (2024) Feasible methods to control starch digestibility: strategies for reducing glycemic index of rice noodles. Trends Food Sci Technol 149:104536

Nagy R, Máthé E, Csapó J et al (2021) Modifying effects of physical processes on starch and dietary fiber content of foodstuffs. Processes 9:17

Obadi M, Xu M (2024) A review of the effects of physical processing techniques on the characteristics of legume starches and their application in low-glycemic index foods. Int J Biol Macromol 279:135124

Oba-Yamamoto C, Takeuchi J, Nakamura A et al (2022) Impact of low-starch high-fiber pasta on postprandial blood glucose. Nutr Metab Cardiovasc Dis 32:487–493

Panlasigui LN, Thompson LU, Juliano BO et al (1991) Rice varieties with similar amylose content differ in starch digestibility and glycemic response in humans. Am J Clin Nutr 54:871–877

Parada J, Aguilera JM (2012) Effect of native crystalline structure of isolated potato starch on gelatinization behavior and consequently on glycemic response. Food Res Int 45:238–243

Pereira EV, de Assis Costa J, de Cássia Gonçalves Alfenas R (2015) Effect of glycemic index on obesity control. Arch Endocrinol Metab 59:245–251

Rahim MS, Mishra A, Katyal M et al (2020) Marker-trait association identified candidate starch biosynthesis pathway genes for starch and amylose–lipid complex gelatinization in wheat (*Triticum aestivum* L.). Euphytica 216:151

Romão B, Falcomer AL, Palos G et al (2021) Glycemic index of gluten-free bread and their main ingredients: a systematic review and meta-analysis. Foods 10:506

Scioli MG, Storti G, D'Amico F et al (2020) Oxidative stress and new pathogenetic mechanisms in endothelial dysfunction: potential diagnostic biomarkers and therapeutic targets. J Clin Med 9:1995

Shobana S, Kokila A, Lakshmipriya N et al (2012) Glycaemic index of three Indian rice varieties. Int J Food Sci Nutr 63:178–183

Srikaeo K, Sangkhiaw J (2014) Effects of amylose and resistant starch on glycaemic index of rice noodles. LWT Food Sci Technol 59:1129–1135

Subaitha ZA, Priyadarshini SR, Yoha KS et al (2024) Impact of post-harvest processing techniques on the glycemic index of millets. Food Chem Adv 4:100636

Swinnen SG, Hoekstra JB, DeVries JH (2009) Insulin therapy for type 2 diabetes. Diabetes Care 32:S253–S259

Tan BL, Norhaizan ME (2024a) Chapter 4: Molecular mechanisms of oxidative stress-induced diseases. In: Nutrients and oxidative stress: biochemistry aspects and pharmacological insights, SpringerBriefs in food, health, and nutrition. Springer, pp 31–42

Tan BL, Norhaizan ME (2024b) Chapter 5: High carbohydrates intake and type 2 diabetes. In: Nutrients and oxidative stress: biochemistry aspects and pharmacological insights, SpringerBriefs in food, health, and nutrition. Springer, pp 43–50

Tan BL, Norhaizan ME, Liew W-P-P (2018) Nutrients and oxidative stress: friend or foe?. Oxid Med Cell Longev 2018, 9719584, 24 pages

Tatovic D, Narendran P, Dayan CM (2023) A perspective on treating type 1 diabetes mellitus before insulin is needed. Nat Rev Endocrinol 19:361–370

Varghese S, Awana M, Mondal D et al (2022) Amylose-amylopectin ratio: comprehensive understanding of structure, physicochemical attributes, and applications of starch. In: Thomas S, Ar A, Jose CC, Thomas B (eds) Handbook of polymers. Springer, Singapore, pp 1–30

Wang Q, Guan C, Luo L et al (2024a) Effect of soybean protein isolate, transglutaminase, and konjac glucomannan on the cooking and eating quality and digestibility of rice noodles. Int J Food Eng 21:129–137

Wang Y, Ou X, Al-Maqtari Q et al (2024b) Evaluation of amylose content: structural and functional properties, analytical techniques, and future prospects. Food Chem X 24:101830

Wee MSM, Henry CJ (2020) Reducing the glycemic impact of carbohydrates on foods and meals: strategies for the food industry and consumers with special focus on Asia. Compr Rev Food Sci Food Saf 19:670–702

World Health Organization (2024) Diabetes. https://www.who.int/health-topics/diabetes?gad_source=1&gclid=CjwKCAjwtNi0BhA1EiwAWZaANJQp1PbXl0NfTTh9SqjzT0Ol0rHt3OZUa1YL8Jsx2tmb0AZNqpbyoRoCq0gQAvD_BwE#tab=tab_1

Zhang L, Hu X, Xu X et al (2011) Slowly digestible starch prepared from rice starches by temperature-cycled retrogradation. Carbohydr Polym 84:970–974

Zhang L, Huang Y, Feng Z et al (2021) Triglycerides promote colorectal cancer cell growth and metastasis through the PI3K/Akt signaling pathway. Oncol Rep 45:705–716

Zhao T, Zhang H, Chen F et al (2022) Study on structural changes of starches with different amylose content during gelatinization process. Starch 74:2100269

Chapter 8
Summary and Future Prospects

Contents

Rice noodles are a gluten-free food and one of the most popular noodle varieties in Asia. They are among the most widely consumed and produced rice-based staple foods globally, particularly in Asian countries such as Korea, Japan, and China. Rice noodles are available in various forms, including fresh, dried, and instant variants, to accommodate diverse culinary preferences and storage needs. They are typically classified based on their shape, thickness, and texture, ranging from thin vermicelli to wide flat noodles. Each type is suited to specific culinary applications, and exhibits the regional and cultural diversity of their use. Traditionally made from polished white rice flour, rice noodles are nutrient-deficient due to the removal of germ and bran during milling. In contrast, brown rice flour, which retains these components, offers higher nutrient content, including fiber, lipids, protein, and minerals, although it exhibits lower cooking quality, such as increased cooking loss and reduced textural firmness (Muniandy and Gannasin 2019).

Starch is the primary component of rice noodles, which significantly impacts their structure and texture. Amylopectin and amylose fractions play critical roles in fresh rice noodle formation, with higher amylopectin content contributing to better texture. However, brown rice flour, due to its lower starch content and retained bran, exhibits reduced paste viscosity, adversely affecting cooking and textural properties (Bhattachrya et al. 1999; Saleh et al. 2014). Germination has emerged as a low-cost technique to enhance the nutritional quality of rice noodles. It activates endogenous enzymes in brown rice, modifying its fiber, protein, and starch content while increasing bioactive compounds like γ-aminobutyric acid (GABA). Germination also reduces resistant starch and bitterness, improving the cooking and sensory properties of brown rice noodles (Chen et al. 2024).

B. L. Tan, M. E. Norhaizan, *Rice Noodles: Quality Improvement, Processing Techniques and Nutritional Aspects*, SpringerBriefs in Food, Health, and Nutrition, https://doi.org/10.1007/978-3-031-98649-9_8

Extrusion technology plays a pivotal role in improving rice noodle quality. By applying heat and mechanical force, extrusion facilitates starch gelatinization, protein denaturation, and fiber solubilization, enhancing the structure and properties of gluten-free rice noodles. Incorporating rice protein during extrusion has been shown to reduce cooking loss and improve textural properties, while extruded rice bran helps delay starch retrogradation during storage (Detchewa et al. 2022; Tian et al. 2022). Functional ingredients such as red lentil protein, potato starch, and defatted coconut flour have been evaluated for their effects on rice noodle properties. Ultrasound- and enzyme-treated red lentil protein enhances protein digestibility and reduces starch digestibility, making rice noodles more nutritious. Fermentation with beneficial microorganisms like *Lactobacillus plantarum* improves the sensory attributes, antioxidant activity, and flavor of rice noodles. Fermented rice bran enhances free phenolic content, reduces aldehydes, and increases alcohols, contributing to better palatability and antioxidant capacity (Yi et al. 2022).

The glycemic index (GI) of rice noodles is an important nutritional consideration, particularly for individuals managing blood glucose levels. Key factors influencing the GI include rice variety, amylose content, and processing methods. Adjustments to production techniques, such as incorporating resistant starches or alternative grains, can effectively lower the GI, making rice noodles more suitable for health-conscious consumers and individuals with metabolic disorders. Overall, rice noodles represent a versatile food product with deep cultural significance and considerable innovation potential. Advances in raw material selection, processing technologies, and nutritional optimization are pivotal to meeting the evolving preferences and demands of consumers.

References

Bhattachrya M, Zee SY, Corke H (1999) Physicochemical properties related to quality of rice noodles. Cereal Chem 76:861–867

Chen Z, Liang N, Zhang H et al (2024) Resistant starch and the gut microbiome: exploring beneficial interactions and dietary impacts. Food Chem X 21:101118

Detchewa P, Prasajak P, Phungamngoen C et al (2022) Substitution of rice flour with rice protein improved quality of gluten-free rice spaghetti processed using single screw extrusion. LWT 153:112512

Muniandy SD, Gannasin SP (2019) Physicochemical and sensory characteristics of brown rice (*Oryza sativa*) noodles substituted with mung bean (*Vigna radiata*) powder. Malays J Nutr 25:351–359

Saleh M, Abuwaar Z, Akash MW et al (2014) Effect of stabilized rice bran fractions on the formation of rice flour pasting properties. Cereal Chem 91:603–609

Tian X-H, Tan B, Wang L-X et al (2022) Effect of rice bran with extrusion cooking on quality and starch retrogradation of fresh brown rice noodles during storage at different temperatures. Cereal Chem 99:1296–1307

Yi C, Xie L, Cao Z et al (2022) Effects of rice bran fermented with *Lactobacillus plantarum* on palatability, volatile profiles, and antioxidant activity of brown rice noodles. Int J Food Sci Technol 57:5048–5056

Conclusion

This book presents a comprehensive review of various aspects of rice noodles, including the types of noodles, processing of rice noodles, raw materials and the roles in rice noodles, factors influencing rice noodles quality, qualities and nutritional values for improvement of rice noodles, and nutritional considerations, such as the glycemic index. The production of high-quality rice noodles and rice-based products is largely influenced by the specific physicochemical properties of the protein components in the rice flour. Other components such as water, salt, oil, hydrocolloids and polyphosphates, preservatives, and functional ingredients also play a significant role in determining the desired characteristics and organoleptic qualities. Hence, optimizing the composition of these constituents, along with the selection of appropriate ingredients and processing techniques, is essential for achieving the optimum quality of rice noodles. Collectively, this book offers a holistic perspective on the elements that shape the sensory, functional, and nutritional characteristics of rice noodles.

Competing Interests

The authors declare that there is no conflict of interest regarding the publication of this paper.

B. L. Tan, M. E. Norhaizan, *Rice Noodles: Quality Improvement, Processing Techniques and Nutritional Aspects*, SpringerBriefs in Food, Health, and Nutrition,
https://doi.org/10.1007/978-3-031-98649-9

Index

B. L. Tan, M. E. Norhaizan, *Rice Noodles: Quality Improvement, Processing Techniques and Nutritional Aspects*, SpringerBriefs in Food, Health, and Nutrition,
https://doi.org/10.1007/978-3-031-98649-9

The manufacturer's authorised representative in the EU is Springer
Nature Customer Service Centre GmbH, Europaplatz 3, 69115 Heidelberg,
Germany. If you have any concerns regarding our products, please
contact ProductSafety@springernature.com

Printed and bound by CPI Group (UK) Ltd, Croydon, CR0 4YY
01/10/2026
02226351-0007